1992

CATHOLIC PERSPECTIVES ON MEDICAL MORALS:
FOUNDATIONAL ISSUES

PHILOSOPHY AND MEDICINE

Editors:

H. TRISTRAM ENGELHARDT, JR.

Center for Ethics, Medicine, and Public Issues,
Baylor College of Medicine, Houston, Texas, U.S.A.

STUART F. SPICKER

School of Medicine, University of Connecticut Health Center,
Farmington, Connecticut, U.S.A.

VOLUME 34

CATHOLIC PERSPECTIVES ON MEDICAL MORALS

Foundational Issues

Edited by

EDMUND D. PELLEGRINO
Kennedy Institute of Ethics, Georgetown University, Washington, D.C.

JOHN P. LANGAN
Kennedy Institute of Ethics and Woodstock Theological Center,
Georgetown University, Washington, D.C.

and

JOHN COLLINS HARVEY
School of Medicine, Georgetown University, Washington, D.C.

KLUWER ACADEMIC PUBLISHERS
DORDRECHT / BOSTON / LONDON

Library of Congress Cataloging in Publication Data

Catholic perspectives on medical morals.
    (Philosophy and medicine; v. 34)
    Includes bibliographies and index.
    1. Medical ethics.   2. medicine—Religious aspects—
Catholic Church.   3. Catholic Church—Doctrines.
I. Pellegrino, Edmund D., 1920–   . II. Langan,
John, 1940–   . III. Harvey, John Collins.
IV. Series.
R725.56.C37   1989   174'.2   88–13159
ISBN 1–55608–083–2

Published by Kluwer Academic Publishers,
P.O. Box 17, 3300 AA Dordrecht, The Netherlands.

Kluwer Academic Publishers incorporates
the publishing programmes of
D. Reidel, Martinus Nijhoff, Dr W. Junk and MTP Press.

Sold and distributed in the U.S.A. and Canada
by Kluwer Academic Publishers,
101 Philip Drive, Norwell, MA 02061, U.S.A.

In all other countries, sold and distributed
by Kluwer Academic Publishers Group,
P.O. Box 322, 3300 AH Dordrecht, The Netherlands.

Printed in The Netherlands

TABLE OF CONTENTS

INTRODUCTION

CATHOLIC PERSPECTIVES AND CONTEMPORARY
MEDICAL MORALS

A Catholic perspective on medical morals antedates the current world-
wide interest in medical and biomedical ethics by many centuries[5].
Discussions about the moral status of the fetus, abortion, contraception,
and sterilization can be found in the writings of the Fathers and Doctors
of the Church. Teachings on various aspects of medical morals were
scattered throughout the penitential books of the early medieval church
and later in more formal treatises when moral theology became recognized as a distinct discipline. Still later, medical morality was incorporated into the many pastoral works on medicine. Finally, in the
contemporary period, works that strictly focus on medical ethics are
produced by Catholic moral theologians who have special interests in
matters medical.

Moreover, this long tradition of teaching has been put into practice in
the medical moral directives governing the operation of hospitals under
Catholic sponsorship. Catholic hospitals were monitored by Ethics
Committees long before such committees were recommended by the
New Jersey Court in the Karen Ann Quinlan case or by the President's
Commission in 1983 ([8, 9]).

Underlying the Catholic moral tradition was the use of the casuistic
method, which since the 17th and 18th centuries was employed by
Catholic moralists to study and resolve concrete clinical ethical dilemmas. The history of casuistry is of renewed interest today when the case
method has become so widely used in the current revival of interest in
medical ethics[11].

Until recently the Catholic perspective was the most influential,
formal, and comprehensive source for what is now called biomedical
ethics. The writings of most physicians outside that tradition were until
that time largely hortatory, and concentrated on professional ethics –
the relationships of physicians to patients and to each other rather than
on the more fundamental philosophical principles undergirding their
practices.[1] Few if any philosophers took any interest in the specific prob-

1

Edmund D. Pellegrino et al. (eds.), Catholic Perspectives on Medical Morals, pp. 1–6.
© *1989 Kluwer Academic Publishers. Printed in the Netherlands.*

lems of medical morality. The field was generally left to the theologians.

Two decades ago all of this changed abruptly. New and unprecedented moral dilemmas emerged from the prodigious advances of medical technology. Moral pluralism became more explicitly the rule in American life than it had ever been before. Public awareness fostered by media reports of controversial and sensational cases grew rapidly. The courts were increasingly called on to settle conflicts among decision makers whose fundamental values differed sharply. Today medical ethics is a topic of everyday discussion, and a constant presence in all the media, in the courts, and in state legislatures.

One of the most significant of the recent changes has been the entry of professional philosophers into the dialogue. Until recently, philosophers had not been engaged in formal analyses of medical ethical issues. They began by teaching in medical schools and university hospitals[7]. Soon they were writing the scholarly papers and treatises that have become the dominant intellectual resources for contemporary medical ethics.

This philosophical turn has changed the whole complexion of medical ethics. The centuries-old theological approach has been superceded. Its normative thrust has given way to secular methodology. A strongly utilitarian, pragmatic, and eclectic bias has replaced the previous deontologic character of medical ethical discourse. These characteristics of philosophical ethics make it more attractive than theological ethics to medical schools and to physicians in a secular, pluralistic society. Its spirit and methodology have dominated the reports of Presidential Commissions, as well as hospital ethics committees, the courts and the media. As a result, theologians – Catholic, Protestant and Jewish – have lost influence in the field even among their own denominational colleagues.

Historically the Catholic perspective placed a heavy emphasis on sexual, reproductive, and marital morality. This arose from the concern of confessors, moral theologians, and the Church for adherence to Catholic teachings respecting marriage, the family and the virtue of chastity. The Church's concerns are heightened today not only by the revolution in sexual morality and marriage that characterizes our times, but also by the availability of a multitude of new reproductive technologies that separate reproduction from the procreative act[10].

Despite this emphasis on sexual morality the Catholic perspective has attended seriously to other important dimensions of biomedical ethics.

The moral obligations of the physician, for example, were topics of serious concern in the 15th-century work of St. Antoninus of Florence ([5], p. 26). They were addressed usually in later works under the rubric of "duties of one's station" in life. More recently the Church has taken leadership in a dimension of biomedical ethics largely neglected elsewhere – the social ethics of medicine, the obligations of society regarding the care of the sick, and of the sick poor ([1, 3, 4, 10, 12]).

But even aside from its history and specific concerns, a Catholic approach presents several fundamental anomalies when considered from the view point of modern-day biomedical ethics.

For one thing, it employs sources of moral insight – scripture, tradition, and authoritative teachings – which are neither central nor normative for a purely philosophical ethics. Its theory, at least in the area of medical ethics, is commonly taken to be deontological. At the same time it relies heavily on Aristotelian teleology and virtue theory. In addition, the metaethical structure of a Catholic perspective is framed by belief in an objective order of morality and a specific philosophical and theological anthropology, which takes into account the spiritual as well as the material destinies of human life. This runs counter to the historicist bias in much of ethics today. A Catholic perspective transcends ethics as a discipline. Its ultimate norm is the law of Charity as enunciated in the Sermon on the Mount. This goes beyond coherent analytical argumentation based on traditional medical ethical principles, though it is not inconsistent with such principles. Finally, the Catholic perspective depends on that special fusion of theology and philosophy that is the distinctive mark of the Catholic intellectual tradition.

The conference which served as the foundation of this volume was convened with several convictions in mind. First of all, there is a pressing need to provide some alternatives to the prevalent tenor of ethical discourse, which is today so largely secular, analytical, and procedural. Substantive issues are often slighted to the discomfiture not only of Catholics but of persons of other religious persuasions as well. There are still large numbers of people for whom the religious sources of ethics are still determinative.

Moreover, there is a need to clarify the Catholic perspective itself, since internal pluralism has grown steadily within post-conciliar Catholic moral theology. Catholic thinkers have entered into dialogue with the mainstream thinkers in secular biomedical ethics, but their contributions are sometimes compromised by internal differences about the place of ecclesiastical authority or debates about what in fact defines *Catholic Christian* ethics.

Finally, the breadth of a Catholic perspective needs to be asserted beyond a narrow concern for sexual and reproductive morality with which it is too frequently misidentified. It is necessary for Catholics as well as non-Catholics to realize that a Catholic perspective also has much to say about what is required of the good physician, nurse, or hospital administrator; about the nature of the healing apostolate; the collective responsibility of the Christian community for the care of the sick; justice in the allocation of scarce resources; responsibility for care of the poor; medical profiteering; and a host of other specific questions about the obligations that flow from a commitment to Christianity.

Obviously one conference can not touch on all these questions. Instead, we have selected a few "foundational" issues which we consider most pertinent at this juncture – (1) the philosophical and theological groundings for a Catholic perspective, (2) the issue of pluralism within and outside the Catholic tradition, (3) the crucial question of the relationship of ethical inquiry to the legitimate teaching authority of the Church, and (4) the place of the virtue of Charity in medical morals[2].

A wide variety of questions is touched on, directly or tangentially, by the contributors to this volume: Does the Catholic perspective entail differences in the content or method of the ethical enterprise? What dimensions does a faith commitment add to the obligations of philosophical ethics? Is the enterprise of ethics as a rational and reasoned discipline compatible with a commitment to Catholic Christianity which must take into account the teachings of an authoritative Church? What is the proper relationship of authority and freedom of inquiry which permits Catholic thinkers to engage new questions and re-examine old ones, in order to deepen our understanding of the obligations that bind Catholic Christian health professionals, institutions, patients, and the entire Catholic community? What differences does a commitment to Christian charity as an ordering principle make to the theory and practice of medical ethics? How can the reality of pluralism, within and outside the Catholic community, be fruitfully and constructively addressed?

This wide range of questions impels us to assemble these essays under the rubric of "medical morality", though we are cognizant of the relative infrequency of this usage. Our intent is to encompass the topics currently included under such designations as 'medical ethics', 'bio-ethics', or 'biomedical ethics' – all of which are subject to examination from a Catholic perspective. The term medical morality also suggests a closer link and continuity with the Catholic tradition's long history of concern with issues in medical ethics as one branch of a broader inquiry into Christian living. Finally, medical morality implies a systematic moral philosophy, a groundwork for the inquiry into the *right* and the *good* in medical activity, the reasons for pursuing and being good, and the kinds of persons health professionals ought to be.

We believe these essays and commentaries qualify literally as a "perspective" – a "seeing through" in two senses: (a) they penetrate to the fundamental questions most pertinent to the committed Christian, and (b) they do so from the viewpoint of a long tradition of medical morality traceable to the primitive church.

Georgetown University   EDMUND D. PELLEGRINO
Washington, D.C.   JOHN LANGAN
                                           JOHN COLLINS HARVEY

NOTE

[1] Exceptions to some degree were the 18th-century physicians: John and James Gregory of Edinburgh and Thomas Percival of Manchester. See [6].

BIBLIOGRAPHY

1. Catholic Health Association: 1986, *No Room in the Marketplace: The Health Care of the Poor*, Catholic Health Association, St. Louis, Missouri.
2. Curran, C. E. and McCormick R. (eds.): 1980, *The Distinctiveness of Christian Ethics: Readings in Moral Theology II*, Paulist Press, New York.
3. Haughey, J. C. (ed.): 1977, *The Faith That Does Justice, Woodstock Studies II*, Paulist Press, New York.
4. John Paul II: 2/11/1984, 'On the Meaning of Human Suffering, Apostolic Letter'. United States Catholic Conference, Washington, D.C.
5. Kelly, D. F.: 1979, *The Emergence of Roman Catholic Medical Ethics in North America: A Historical – Methodological – Bibliographical Study*, The Edward Mellen Press, New York.
6. Pellegrino, E. D.: 1986, 'Percival's Medical Ethics: The Moral Philosophy of an 18th Century English Gentleman', *Archives of Internal Medicine* **146**, 2265–2269.

7. Pellegrino, E. D. and McElhinney, T. K.: 1980, *Teaching Ethics, the Humanities, and Human Values in Medical Schools: A Ten-Year Overview*. Society for Health and Human Values, Washington, D.C.
8. President's Commission for the Study of Ethical Problems in Medicine and Biomedical and Behavioral Research: 1983, *Deciding to Forego Life-Sustaining Treatment: A Report on the Ethical, Medical and Legal Issues in Treatment Decisions*, United States Government Printing Office, Washington, D.C.
9. *In re Quinlan*: 1976, 70 NJ 10; 355 A2d 647.
10. The Sacred Congregation for the Doctrine of the Faith: 1987, *Instruction on Respect for Human Life in its Origin and on the Dignity of Procreation*, Vatican City.
11. Toulmin, S. and Jonsen, A. R.: 1988, *The Abuse of Casuistry: A History of Moral Reasoning*, University of California Press, Los Angeles.
12. U. S. Catholic Conference: 1981, 'A Pastoral Letter of the American Bishops', United States Catholic Conference, Washington, D.C.

ACKNOWLEDGMENT

The editors wish to express their gratitude to the Raskob Foundation, an anonymous donor, and Georgetown University for their financial support of the Twenty-Fourth Trans-Disciplinary Symposium on Philosophy and Medicine [October 13–16, 1986] and the preparation of this volume.

PART I

PROLOGUE

CARLO MARIA CARDINAL MARTINI

SOME BASIC CONSIDERATIONS ON MORAL TEACHING IN THE CHURCH

As a general introduction, I would like to propose some reflections on two points. The first is about the development of bioethical research and the emergence every day of a more complex ethical problematic. In this context I shall emphasize the mutual enrichment which has been brought out from the dialogue between the biological sciences (biology, genetics, medicine), ethics in general, and applied ethics in particular. The second point will relate to social and juridical implications of bioethical problems, that is, the importance in this historical moment, of the dialogue between morality and law, and of the presence of Christian attitudes in public life. By way of introduction I shall recall briefly some of the major changes brought about by the second Vatican Council on the matter.

I. CHANGES BROUGHT ABOUT BY VATICAN II

In my summary exposition I shall proceed from the new emphases which were introduced into moral thinking by Vatican II, and which have fostered a development and a deepening of moral teaching in the last 25 years.

The Vatican II Council "Optatam totius" decree recommended that: "Special attention needs to be given to the development of moral theology. Its scientific exposition should be more thoroughly nourished by scriptural teaching. It should show the nobility of the Christian vocation of the faithful, and their obligation to bring forth fruit in charity for the life of the world" ([3], 16). With this recommendation it summed up the aspirations of many theologians who in their teachings were already trying to find a unifying principle that would give cohesion to the foundations of Christian Morals, beyond the extrinsic character and the casuistry of many pre-council textbooks.

And here I remember Hirscher, one of the first in the XIXth Century renovation movement, in whose view Christian Morals should show the way towards the realization of the Kingdom of Christ, a spiritual kingdom where the fundamental Constitution would be the doctrine of

9

Edmund D. Pellegrino et al. (eds.), Catholic Perspectives on Medical Morals, pp. 9–19.
© *1989 Kluwer Academic Publishers. Printed in the Netherlands.*

the Sermon on the Mount; Tillman, who would center his moral teachings on the following of Christ; Mersch, who would try to do the same in
the light of the doctrine of the Mystical Body; and finally, Bernard
Häring, whose work "The Law of Christ" had great impact in the years
before the Council. In this work Häring focussed on the call-response
dialectic: God calls Man in Christ, gives him the possibility of answering
Him, which is made concrete in the reality of a Christian vocation.

In addition to this emphasis on the ethical dimension of the faith,
other essential aspects of moral theology were also worked on. Namely,
the question of the relationship between conscience and the moral
norm, the morals of attitudes with a renovation of the treatment of
human acts and of the interpretation of the morals of sin and of
conversion, of habits and of virtues, the specifity of Christian Morals,
the relationship between the Magisterium and the theologians.

The special 1985 Synod explicitly recommends a closer communication and a mutual dialogue between *bishops and theologians* for the
edification and a deeper understanding of the faith.

There is no doubt whatsoever that some can feel tempted, when
seeing the road still left ahead, to underestimate what has already been
done. But it is important to underline the importance of the changes
that have occurred.

Vatican II, as it was clearly emphasized in the Extraordinary Synod
for Bishops of 1985, is an irreversible step forward in the relationship
between the Church and modern society.

I shall mention explicitly *Gaudium et spes* where the problem of the
relationship of the Church with modern man is most explicitly
addressed.

In *Gaudium et spes* the Church confesses that it shares "the pleasures
and hopes, sorrows and anxieties of the men and women of our time",
where it declares that it "wants only one thing: under the guidance of
the Holy Spirit, to continue the very work of Chirst who came to this
world to give witness to the truth, to serve and not to be served", and
that "to achieve this goal it is the Church's constant duty to examine the
signs of times closely and to interpret them in the light of the
Gospel . . ." [2].

The transformation which all of this implies should be measured by its
depth rather than its speed. To compare it to the rapidity of change in
modern technologies, such as computer science and telecommunications, which make great social transformations possible, is not entirely

appropriate because the changes in the field of philosophy/theology are the ones that ultimately enable us to direct these technological changes and give them an orientation consistent with the dignity of the human person.

But I have still to mention one main point that shows the relation between the Council and the moral teaching of the Church.

The Council has been called "pastoral". This expression has of course different meanings and has been explained in many ways. The question is, how does this apply to the moral teaching of the Church, and especially to the concern of a Bishop on moral questions. I find it very important that the moral teaching given in the Church should duly preserve and emphasize what the New Testament calls "*Pareneses*" or Exhortation. There are many instances of this in the pages of the New Testament that speak of ethical problems. Let us take, for example, *Ephesians* 4:25–6:20 and similar passages. The point is here not exactly to decide what is right or wrong in a given situation, but to foster a way of acting according to certain lines that reproduce the way of life of Christ and of God himself. "Therefore be imitators of God, as beloved children. And walk in love, as Christ loved us . . . (*Eph* 5:1–2).

What strikes us in this "moral section" of the Epistles is the emphasis on exhortation and encouragement, more than the strict definition of what is acceptable and what has to be avoided. Or better, what is to be done or to be avoided is seen as a part of a process which leads to Christian fulfillment, to a way of life which is Christian sanctity.

We can gather from this that one of the main functions of the moral preaching in the New Testament is to help the faithful to walk towards the goal of a life very high in moral quality, which is understood as an "imitation of God", an "imitation of Christ, and "perfection", and which constitutes the essence of the kingdom of God, of the heavenly Jerusalem.

The main concern is not to decide in a static sense where lies the boundary between good and evil, between what is permitted and what is prohibited; the goal is not to create perfect moralists, but to help people to walk towards Christian sanctity, towards a clear ideal of moral perfection.

From that it follows that the moral life of the Christian is a path, a way, towards a very high achievement, which is to live as Christ did, to imitate his freedom of heart, his love for the poor, to be chaste, to be kind and compassionate. The main concern is then not simply "morality" in the usual sense of the term, but holiness, fullness of Christian life.

This has at least two major consequences. First of all, in the face of a new problem or a new moral question, it will not be enough to answer the question, "is this feasible? is this permitted?", but the question: "how does this or that behavior, this or that doctrine promote fullness of Christian life, personally and socially, for the single conscience and for the entire people of God?"

Secondly, important words never to be forgotten are "more", "perfection", and "sanctity". It is not enough to be satisfied with behavior that does not harm. The question will often be: what is more conductive to Christian fullness? what is the ultimate goal of human perfection to be achieved? In the frame of this question lies the right attitude which helps to tackle more frankly some new difficult moral questions. It may also happen that there will not be a clear, definite answer to a particular problem, because it is still in the process of definition in its nature and as yet unable to be judged with complete certainty. But also in contexts like these it will always be necessary to point out where lies the goal of Christian maturity to be attained and which direction to follow to come to such point of Christian perfection.

Of course the specialist in ethics, confronted with new problems, will be concerned about a clear judgement to be reached according to right principles of moral decision. In this symposium you will treat some of these principles, and I shall gladly listen to the different propositions. But the pastor will be even more concerned about the question of how much through a certain behavior the Christian community will be helped to grow in love and hope.

II. BIOMEDICAL ADVANCES AND MORAL DIALOGUE

The biomedical advances that have been occurring during this last quarter of our century are to be seen in the general framework of other scientific and technological advances. Today we see scientific progress as a huge sociocultural phenomenon that has led to an external mode of projection having a direct influence on our ways of life, and an indirect influence on our representations and our systems of values. *Technology* represents the visible face of modern science to which it is intimately linked, to the point (or the danger) of becoming identical.

When we consider the advances made in the fields of physics (nuclear energy), of microelectronics (computer science, telecommunications) and of molecular and genetic biology (the possibilities of DNA recombinations), and specifically the techniques known as genetic engineering, we see that they have *a radical and totally encompassing* nature. As with any type of discovery, these advances have an *ambivalent* character. They can be used for good or for evil, but, more than other types of technology, they do *radicalize the effects of human actions*, for they lead human beings to a deep knowledge and deep power over things and over each other. Apart from being radical, their action is totally encompassing. By this is meant that these technologies extend their action *over the whole world in a matter of years*, thus influencing all aspects of human life. These technologies also introduce major risks of imbalance and structural maladjustments of all types: economic, social, and political. The combination of knowledge and power tends to concentrate itself in fewer and fewer hands and the powers of this generation are increasingly able to drastically reduce those of the next generation.

It is therefore absolutely essential to prevent technological dynamics from inverting the *relation between the means and the end*, which would severely endanger the elements of responsibility at the level of the individual conscience and, above all, at the level of the collective powers.

Let us not forget that science as well as technique, geared towards transforming the world, finds its justification in the service it gives man and mankind and, ultimately, to the basic meaning of life and of human activity. I will recall here John Paul II's addresses to scientists and to a group of Nobel prize winners in 1980: To scientists: " . . . There is no reason why we should see our technical and scientific culture as opposed to a world created by God. Technical know-how can clearly be used for good as well as for evil. The person investigating the effects of poisons will be able to use that knowledge to cure as well as to kill. But there should be no doubt as to the direction towards which we must go in order to distinguish good from evil. Technical science, geared towards the transformation of the world, finds its justification in the service it gives to man and mankind . . ."[6].[1]

To Nobel prize winners: " . . . Is it surprising to start hearing about a legitimacy crisis in science and even of a crisis affecting the orientation of our scientific culture as a whole? Science alone is unable to give a comprehensive answer to the problem of the basic meaning of human life and activity. This meaning is revealed when reason, going beyond physical data, turns to metaphysical methods to achieve a contemplation of the 'final causes' and so unveils the ultimate explanations that are able to shed light on human events and give them a meaning.

The quest for the final meaning is of a complex nature and exposed to the risk of error, and often enough man would go on groping in darkness were he not helped by the light of faith. Christian revelation contributed invaluably to the conscience that modern man could gain regarding his own dignity and rights. I do not hesitate to repeat here what I told the members of UNESCO: "The statements regarding man belong as a whole to the very substance of Christ's message and of the mission of the Church, in spite of whatever declarations critical minds might have made in this respect . . ." [5].[2]

It would be unfair to ignore *the scientific community's own self-criticism*. Today more than ever science itself is questioning the assumption that the method of physics is the model for all forms of valid knowledge; also questioned is the presumed absolute objectivity of scientific discourse, and no one believes any longer in the straight dichotomy between belief and science. There are also voices which demand philosophical reflection as a complement to scientific thinking. These voices are in frank contrast with the positivistic inclination to place the human sciences within the sciences of nature.

The *non-clinical physician* may be still in the position to be tempted to share the positivistic approach and to hold that truth is equivalent to scientific truth, and therefore to act in his/her works and research without moral justification for research objectives.

The *clinician* – working closer to suffering, pain and death – is constantly reminded of his/her own contingency and finiteness. Today both have an increased power over life in all its aspects, from the laboratory to death, and the consequences of their actions are more serious and lasting than in other times. When we think of the possibilities of modifying the human gene, we feel our whole being shiver and the word we want to utter is "caveat!". Biomedical dialogue is indispensable in that common perception of a responsibility that affects the patients and can, as in other scientific decisions, affect other future generations. Such a dialogue must look deeper into the very foundations, if it is to avoid the limitations of casuistry. Its mission is to help us understand that the human species is risking its destiny, not in the naked objectivity of science-technology, but in subjective inwardness, in its responsible freedom, in other words, in its very conscience. The help the Church can receive from the dialogue with Medicine and with science generally is reflected in these words from John Paul II: " . . . We should add that the Church recognizes with gratitude all that she owes to research and science. I had the occasion to say this to the Pontifical Council for Culture on 18 January 1983: 'Let us think of how the results of scientific research help us to know the universe better, to understand better the mystery of man. Think of the advantages which the means of communication and contact among peoples offer to society and to the Church; let us think of the ability to produce incalculable economic and cultural wealth, and especially to promote the education of the masses, and to cure diseases previously thought incurable. What admirable achievements! All this is to man's credit. And all this has greatly benefitted the Church itself, in its life, its organization, its work and its own activity.' And if we address ourselves now more directly to the scientific world, does one not see today how the *greater sensitivity* of scholars and researchers to spiritual and moral values brings to your disciplines a new dimension and a more generous openness to what is universal? This attitude has greatly facilitated and enriched the dialogue between science and the Church" ([7], p. 7).

In a particular manner, the clinician has not only come to an aware-ness of the great responsibility which accompanies his/her real power, but has also been pleasantly surprised by the interest which moralists have shown for the moral problems raised by medical decisions. Para-doxically, it must be said that, thanks to this particular field of medical ethics, case analysis, once so vilified by morals has become the accepted analytical method. Toulmin in a 1982 article expressed this with great clarity: "How did the fresh attention that philosophers began paying to the ethics of medicine, beginning around 1960, move the ethical debate beyond this standoff? It did so in four different ways. In the place of the earlier concern with attitudes, feelings and wishes it substituted a new preoccupation with situations, needs and interests; it required writers on applied ethics to go beyond the discussion of general principles and rules to a more scrupulous analysis of the particular kinds of "cases" in which they find their application; it redirected that analysis to the professional enterprises within which so many human tasks and duties typically arise; and finally, it pointed philosophers back to the ideas of "equity", "reasonableness", and "human relationships", which played central roles in the *Ethics* of Aristotle but subsequently dropped out of sight (5, esp. 5.10.1136b30–1137b32). . . . In traditional case morality, as in medical practice, the first indispensable step is to assemble a rich enough 'case history'. Until that has been done, the wise physician will suspend judgement. If he is too quick to let theoretical considerations influence his clinical analysis, they may prejudice the collection of a full and accurate case record and so distract him/her from what later turn out to have been crucial clues. Nor would this outcome have been any surprise to Aristotle either. Ethics and clinical medicine are both prime examples of the concrete fields of thought and reasoning in which (as he insisted) the theoretical rigor of geometrical argument is unattainable: fields in which we should above all strive to be reasonable rather than insisting on a kind of exactness that "the nature of the case" does not allow (5, 1.3.1094b12–27)" ([10], pp. 737 & 742).

The openness of scientists, and in particular of physicists, biologists, and physicians to the moral dialogue in the search for the values that should shape our humaneness, coincides with the historical moment when the Church has recognized the autonomy of wordly realities (*Gaudium et spes*, 36) and the right of the human person to religious freedom (*Dignitatis humanae*). The dialogue between Faith and Science is more urgent than ever and necessary before. At the same time, it receives a better reception both from Science and from the Church, as compared with other periods in History.

III. MEDICAL ETHICS AND PUBLIC POLICY: THE PROBLEM OF CHRISTIAN PRESENCE

I notice that the problems of Medical Ethics refer to three main areas: (1) the problems relating to the beginning and the end of life; (2) the problems of the autonomy of the patients (patients' rights); and (3) the problems related to the right distribution of limited resources (macroeconomic and microeconomic levels). Underlying all of these areas, there is always a problem which requires great attention and deeper consideration. It is not a new problem. It comes from classical antiquity, but has acquired a vital importance in our times. It is the problem of the relationship between Ethics and Law in a pluralistic society.

This problem has come to the forefront also in the field of medicine in view of the following facts: medicine has the power to affect life itself and the quality of life; the health sector has growing needs and expectations; the cost of health services is continually increasing. All this results in complicated problems when new legislation is considered. Such legislation must take into account the diversity of the values that deserve protection.

I think that this problem has not yet been tackled with due attention, although, especially in the United States (I think in particular of the contributions of J. Courtney Murray, R. McCormick and J. Bryan Hehir), there have been brilliant efforts to address this problem.

It is necessary to go deeper in order to be able to give a guiding and clarifying answer that could assess the complex web of relations, influences, pressures, and manipulations of information which shape public opinion and to a large extent determine the common "ethos". This pervasive and permissive influence of the social communication media is the reason why in different countries it may be some times necessary, given the circumstances, to turn to the "doctrine of the moral minimum" (moral values necessarily under the protection of the law in order to safeguard social order). Certainly, a Christian must always and everywhere be the salt of the earth and the light of the world, but how is (s)he to have an efficient influence on public life? Must we simply accept a society which will forego the growth of real human progress toward the goals which God set for man?

I will leave for you the difficult task of tackling such delicate matters, and will limit myself to recall some statements of well-known authors, which help to formulate more clearly the problem.

The first statement is about the "possibility" and the "effectiveness" of legislation. What morality ought we to legislate? In a pluralistic society the answer to that question is not easy. Morality can translate into public policy only if it survives the test of feasibility or, to use the expression of John Courtney Murray, S. J., of "possibility". He wrote: "A moral condemnation regards only the evil itself, in itself. A legal ban on an evil must consider what St. Thomas calls its own "possibility". That is, will the ban be obeyed, at least by the generality? Is it enforceable against the disobedient? Is it prudent to undertake the enforcement of this or that ban, in view of the possibility of harmful effects in other areas of social life? Is the instrumentality of coercive law a good means for the eradication of this or that social vice? And since a means is not a good means if it fails to work in most cases, what are the lessons of experience in this matter" ([9], pp. 166–167). One of the key ingredients in a policy's feasibility or possibility is a sufficiently broad consensus about the moral character of the act in question. The Rev. J.

Bryan Hehir recently noted that Father Murray's question, "Can we fashion a public consensus to direct and control the purpose and uses of the awesome power of this society?", is still with us. Without such consensus, legal bans will be out of place[8].

The second statement is about the effort required by the Church to express clearly and firmly its social dimension in the public life.

"The social dimension of the Church requires a coordinated effort by the Hierarchy and by the faithful – at his/her own level – to build the web of relations, legal and of all types, which constitute the supernatural common good, i.e., that body of norms and possibilities which is to enable primarily all Catholics and, in some fashion, all the human species to obtain the most perfect realisation of its supernatural destiny. Similarly, it is the responsibility of the civil society – authorities and citizens at the national and international levels – to promote the natural common good, that is, the sum of those conditions of social life which allow social groups and their individual members relatively thorough and ready access to their own fulfillment" ([2], 26).

It is not by hiding or silencing ethical values that one promotes a more human and just order, but rather by making all citizens more sensitive to these values and also the same political power so that this power will find itself forced to promulgate the decision which, here and now, will protect and promote, "in the adequate order of priorities, the largest number of moral values which are at stake"[4].

I will now conclude by remembering André Hellegers, founder of the Kennedy Institute, who had the intuition to combine in one brotherhood scientific and theological knowledge, the preoccupation for immediate problems brought about by biomedical advances and the wider problems of universal repercussions, especially in the less developed countries.

May the Lord enlighten your work.

Archdiocese of Milan
Milan, Italy

NOTES

[1]" . . . Il n'y aucune raison pour regarder la culture technico-scientifique comme opposée à un monde créé par Dieu. Il est assez clair que la connaissance technique peut être utilisée pour le bien comme pour le mal. Celui qui fait des recherches sur les effets des

poisons pourra utiliser cette connaissance pour guérir comme aussi pour tuer. Mais il ne peut y avoir de doute touchant la direction vers laquelle il faut marcher pour distinguer le bien du mal. La science technique, orientée vers la transformation du monde, trouve sa justification dans le service qu'elle rend à l'homme et à l'humanité . . ."

[2]" . . .Faut-il s'étonner si l'on commence aujourd'hui à parler d'une crise de légitimité de la science et même d'une crise touchant l'orientation de l'ensemble de notre culture scientifique? La science à elle seule est incapable d'apporter une réponse complète au problème de la signification fondamentale de la vie et de l'activité humaine. Cette signification se révèle lorsque la raison, dépassant les données physiques, recourt à des méthodes métaphysiques pour arriver à la contemplation des "causes finales" et y découvre les explications suprêmes qui peuvent éclairer les événements humains et leur donner un sens. " . . . La quête de la signification finale est complexe par nature et exposée au danger de l'erreur, et l'homme bien souvent continuerait de chercher à tâtons dans les ténèbres s'il n'était aidé par la lumière de la foi. La révélation chrétienne a apporté une inestimable contribution à la conscience que l'homme moderne a pu acquérir de sa prope dignité et des ses propes droits. Je n'hésite pas a répéter ici ce que j'ai dit aux membres de l'UNESCO: "L'ensemble des affirmations concernant l'homme appartient à la substance même du message du Christ et de la mission de l'Eglise, malgré tout ce que les esprits critiques ont pu déclarer en la matière . . ."

BIBLIOGRAPHY

1. Abbott, W., S. J. (ed.): 1962, *Dignitatis Humanae*, in *The Documents of Vatican II*, Geoffrey Chapman, London.
2. Abbott, W., S. J. (ed.): 1962, *Gaudium et Spes*, in *The Documents of Vatican II*, Geoffrey Chapman, London.
3. Abbott, W., S. J. (ed.): 1962, *Optatum Totius* in *The Documents of Vatican II*, Geoffrey Chapman, London.
4. Cuyas, M.: 1978, 'La Iglesia ante una ley sobre el aborto', *Raison y Fe* **198**, 175–86.
5. John Paul II: 1980, Address to Nobel Prize Winners, December 22.
6. John Paul II: 1980, Address to Scientists, November 15.
7. John Paul II: 1983, *Osservatore Romano*, May 30.
8. McCormick, R.: 1985, 'Therapy or Tampering? The Ethics of Reproductive Technology', *America* **153**, 397–398.
9. Murray, J. C.: 1960, *We Hold These Truths*, Sheed and Ward, New York.
10. Toulmin, S.: 1982, 'How Medicine Saved the Life of Ethics', *Perspectives in Biology and Medicine* **25**, 736–750.

PART II

THE PHILOSOPHICAL FOUNDATIONS

WILLIAM A. WALLACE

NATURE AND HUMAN NATURE AS THE NORM IN MEDICAL ETHICS

A "Catholic perspective on medical morals" should be based in some way on natural law, this being the central theme that underlies most Catholic thought on ethical matters. The expression "natural law," however, can easily be taken to mean something in the human mind that has no clearly identifiable correlate in the ontological order. In this conference devoted to "foundational issues" I thought it well to redress this impression at the outset. Rather than frame medical ethics in a natural-law context, I focus instead on nature and inquire how it, and more particularly human nature, can function as a norm in medical morals.

Let me begin by specifying the meaning of the terms in my title so as to provide a general overview of what I intend and do not intend to do in this paper. "Medical ethics" I take to have the same meaning as bioethics, a specialization within the field of ethics that considers the moral character of acts studied in the health-care disciplines, usually not ascertainable on the basis of common sense and ordinary experience alone but requiring detailed biological and medical knowledge as well. "Ethics" for me is the same as moral philosophy or moral science; I understand this in the Aristotelian-Thomistic sense as a practical science, one that is able to provide norms for human action essentially on the basis of reason alone.[1] My primary concern is thus not moral theology, for the "norm" in my title is a *regula* provided by *recta ratio* and not by any divine or ecclesiastical authority, although I hope that my findings will be applicable within theology also. "Nature" I take in Aristotle's primary sense, from the second book of his *Physics*: an internal substantial principle of characteristic activity, of motion and of rest, on the basis of which we may say that things have "natures" and act and react in conformity with them. (*Physics*, 1:191b 21–23) "Human nature" is then a further specification: it is the type of nature possessed by the rational, voluntary being known as *homo sapiens* or *animal rationale*, whose actions are judged good or bad depending on whether they conform or do not conform to his rational nature.

In light of these definitions, ethics as a practical philosophy, and

23

Edmund D. Pellegrino et al. (eds.), Catholic Perspectives on Medical Morals, pp. 23–53.
© *1989 Kluwer Academic Publishers. Printed in the Netherlands.*

medical ethics along with it, may be presumed to be rooted in a speculative philosophy known as the philosophy of human nature. Again, this philosophy of human nature is itself but a specialization within the philosophy of nature, or natural philosophy.[2] Now the philosophy of nature happens to be my field of study. It has been much neglected in recent Catholic thought, even during the Thomistic revival earlier in this century, partly because of the "metaphysical imperialism" that has dominated Thomism, partly because it has been displaced by a related discipline, the philosophy of science, which happens to be my field of study also.[3] On the basis of my knowledge of these disciplines there are at least two points I wish to make in this paper. The first is that there is a paucity of information about nature and human nature on which to base reliable norms for medical ethics; and the second, that part of this situation at least is traceable to the undeveloped state of our philosophy of nature, notably in its role as a philosophy of science, and so in its failure to keep up with progress in modern science. My major concern in what follows will therefore be to renew the philosophy of nature by assimilating within it a number of developments from recent science that may provide a new foundational perspective on medical morals.

RENEWING NATURAL PHILOSOPHY

Natural philosophy, I would maintain, reached the highest state of its development at the end of the sixteenth century, mainly at the University of Padua through the work of Jacopo Zabarella, Galileo Galilei, and William Harvey, and also at the Collegio Romano, through the efforts of Christopher Clavius and his Jesuit colleagues and disciples.[4] This was done within a tradition that is broadly identifiable as Aristotelian, but focused through Greek, Arab, and Latin commentators, not the least of whom was St. Thomas Aquinas. The part of the *corpus aristotelicum* that was involved included the *Physics, De caelo, De generatione et corruptione*, and *De anima*, plus a number of smaller works known collectively as the *parva naturalia*. It is in these writings that one will find the most clearly articulated analyses of nature and human nature. But, because of the time frame, these analyses were developed in the context of a geocentric universe with finite dimensions, every one of whose heavenly bodies was directly visible to the naked eye, whose duration was less than six thousand years, etc. This was also a universe composed of four elements – fire, air, water, and earth – with a fixed number of species of minerals, plants, and animals that had existed unaltered from the "six days" of God's creative activity. The summit of that creation was Adam, formed directly from the slime of the earth. When God breathed life into the first man, he became the meeting place of spirit and matter, partly angelic, partly terrestrial in his being. His body was composed of humors, ruled by the principles of Galenic medicine, and so on. But his soul was God's direct creation, endowing him with an intellect and a will, in virtue of which he could truly be said to be formed in the image and likeness of his Maker. This was the setting, you must recognize, for Aquinas's famous *Summa Theologiae*, on which much of the Church's moral theology still rests to the present day.

I mention this not to denigrate the natural philosophy that has been bequeathed to us in the Aristotelian-Thomistic tradition, but rather to emphasize the singular difficulty of renovating that philosophy following the scientific revolution of the seventeenth century. Anyone who would attempt it must be thoroughly conversant with the Latin texts in which it had been formulated at the highest stage of its development in the Renaissance, and be able to extract from them their essential content. Just as important, one must know the history of science down to the present, and be able to evaluate it critically, to judge what has been transient and ephemeral and what will retain lasting importance for future ages. And the person who embarks on this enterprise must be prepared to be contradicted almost at every turn along the way. In a field as vast as this, requiring as it does interdisciplinary skills of the highest order, there is no consensus on even the simplest matters.

The task I have set for myself is thus formidable, indeed impossible if I do not claim for myself a number of indulgences. The first of these is that I presume a basic knowledge of natural philosophy in the Aristotelian-Thomistic tradition, and this so that I can employ its technical terminology and presume assent, without proof or extended explanation, to its central theses.[5] The second relates to the philosophy of science, which, as you may know, is in a chaotic state in the present day. I presume a basic sympathy for a realist philosophy of science that takes causality seriously and does not see the problem of induction as an insuperable barrier to valid generalizations.[6] Thus I take truth and certitude as the ultimate goal of scientific inquiry, even though the vast majority of the findings of contemporary science may be regarded as

144,727

fallible and revisable. The third and final indulgence relates to my on-going work on the concept of nature. This requires study, is difficult to understand, and has not yet been subjected to critical evaluation in the literature.[7] Nonetheless, I presume that my views are basically correct and can be built upon to renew the philosophy of nature along the lines indicated.

With this preamble, I first propose to examine a number of problems associated with the causality of nature; as a partial solution to these I offer several models of nature and natural kinds that are open to later development. I then apply these in two areas: in the first I consider the entire order of nature to discuss creation and evolution, from the "big bang" to hominization; in the second, I examine some mechanisms of natural generation, from chemical synthesis and radioactivity, through plant and animal generation, to human procreation. Finally, on the basis of these explorations I shall make some tentative suggestions as to how natural norms may be made available for application in medical ethics.

THE CAUSALITY OF NATURE

Aristotle defined nature as the "principle and cause" of all natural activity and reactivity, but in his many writings he gave few indications of the categories of cause he had in mind in this definition. By categories of cause I mean the four basic kinds: material, formal, efficient, and final. The first two of these, material and formal, are usually called intrinsic causes, and the last two, efficient and final, extrinsic causes. Nature, being an internal principle, may be presumed to involve intrinsic causes, and in fact Aristotle does so identify them in the second book of his *Physics*. (*Physics*, 2:194 a10-b15) This is the basis for his famous hylomorphic doctrine, namely, that every natural substance is internally composed of a basic substrate, *hule*, and a specifying form, *morphe*, that endows it with a particular nature. Indeed, Aristotle allows that both matter and form may be referred to as "nature," while arguing that form, meaning by this what the scholastics, St. Thomas among them, would call *forma substantialis* or substantial form, is the primary referent of this term.

The problem of nature's causality does not lie here; rather it is found when we pass to extrinsic causes and inquire whether nature is an efficient cause of natural activities. This became a serious problem for medievals, including St. Thomas, because of their views of the inorganic world and particularly their teaching on the elements.[8] So as to differentiate the living from the non-living, they would frequently maintain that living things are capable of immanent activities, i.e., life functions that originate from within. Because of this their forms, which are commonly called "souls,"[9] can be regarded as efficient causes, though they do not act as agents directly but rather through the intermediary of active powers with which they are endowed. Non-living things, however, and especially the four elements (which were thought to have a completely homogeneous structure), had no capability for such immanent agency, and thus had to be moved totally from without. Here it was common to invoke Aristotle's principle from the last two books of the *Physics*: *omne quod movetur ab alio movetur*, i.e., whatever is moved is moved by another.[10] One of the most interesting chapters in the history of late medieval philosophy is concerned with the problem of what moves a heavy body in free fall and what moves a projectile after it leaves the hand of the thrower. Gravitational motion was regarded as a natural motion, but there was great resistance to attributing it to a *vis motrix* or *vis gravitatis* within the body, and its efficient cause was usually seen as the generator of the body or whatever removed impediments to its fall, the *removens prohibens*, and so accidentally set it on its motion. Projectile motion, on the other hand, was regarded as a forced or violent motion, and its agent was identified in two opposing ways. Conservative Aristotelians said it was the medium through which the projectile moved, which continued to impel it throughout its motion, whereas their progressive counterparts, departing from the text of Aristotle, attributed it to an *impetus* or *virtus impressa* that moved it from within, albeit in a transient fashion. They were then faced with explaining this peculiar *virtus*, which seemed to have life-like characteristics, and how it could pertain to the inorganic realm.

With regard to final causality, there never was any doubt within the Aristotelian traditon that nature acts for an end; the problem was one of understanding the mechanisms through which it did so. In the *Physics* Aristotle had maintained that the natural philosopher demonstrates through all four causes, and especially through the final cause (*Physics*, II: 7-9). In human affairs, of course, it was readily seen that the end is a most potent cause, the *causa causarum*, first in intention and last in execution. In living things also, the end of generation would seem to be the form that is finally achieved. So St. Thomas would frequently observe that the form of man functions as a cause in a threefold way: as substantial form informing the matter of which he is made; as efficient cause, when he begets another human being; and as final cause, when the same specific form is produced in the offspring and terminates the procreative process.[11] But here again the problem reasserted itself with the inorganic. If a falling body seeks the center of earth as its *terminus ad quem*, how does it do so? What agent can move it so unerringly to its goal, the final cause of its motion?

This brings me to my last vexing problem relating to the causality of nature, that, namely, of the role of the Author of Nature in nature's activities. By Author of Nature I mean the *primum movens immobile* of Aristotle, the *primum esse subsistens* of Aquinas, and generally the God of the Christian and Islamic worlds. Perhaps I should include along with God the "intelligences," the separated substances or immaterial agents of Aristotle and Averroes. These were thought necessary to move the heavenly bodies, and through them, even to induce changes in the sublunary sphere. Well known is St. Thomas's *quinta via* or fifth way, where he argues that non-intelligent agents achieve their ends because directed to them by a Supreme Intelligence, which he identified with the God of creation. Both he and his mentor, St. Albert the Great, were convinced of the adage, *opus naturae est opus intelligentiae*, the work of nature is the work of intelligence ([16], pp.441-463). But whereas Albert was willing to countenance the existence of Aristotle's separated substances, Aquinas rejected that teaching and substituted the angels of revelation instead ([18], p.143-175). For him there was a simple accounting for the marvels of nature. All natural beings are directed to their ends by the Author of Nature, who uses the angelic hosts as his agents in filling out the divine plan. And finally, combining this teleology with God's efficient causality, St. Thomas could develop his teaching on the divine *concursus*: *Deus operatur in omni operante*, that is, God operates directly in every creaturely activity here below. (*Summa Theologiae*, la:q.105,a.5). No problem explaining the fall of a stone or the building of a spider's web: both are from nature, but ultimately the action of the First Cause sustains and directs each in being and in operation. One need not know all the secondary causes involved to be sure of this, and so metaphysics and theology can flourish despite obvious deficiencies in the science of nature.

THE MODELING OF NATURE

Our concern, however, is the science of nature, and we can (and indeed must) go considerably beyond the understanding of nature's causality I have just sketched, which was dominant from the thirteenth to the sixteenth centuries. To do so we must avail ourselves of the considerable information made available by modern science, particularly in the realm of the non-living. I propose to do so in very abbreviated fashion by making use of models. These are not iconic models, in the sense that they picture nature's operation in the way the Bohr model of the atom, say, enables one to visualize a physical or chemical transformation. Rather they are schematic models, basically diagrams that compress a large amount of information in a small space. These can be explicated with the aid of iconic models, but to do that would take me far beyond the aims of this essay.[12]

My first model is taken from scholastic natural philosophy and simply diagrams the constituents of the natural body as these might be understood by a contemporary Thomist (Figure 1). There is nothing controversial about this; it is intended simply to anchor us in the tradition and enable us to make further precisions relating to formal-material causality. A natural body, as shown here, is composed of substance and accidents, the latter pertaining to nine different categories as first schematized by Artistotle. Its substance, diagrammed as the interior core, is itself composed, the two essential components being identified by the letters PM and SF. PM stands for protomatter, the prime matter or *materia prima* of the scholastics, and SF for substantial form, the *forma substantialis* already referred to. In addition to having SF stand for substantial form, however, I wish also to have it stand for "specifying form" and "stabilizing form." My reason for this is that both PM and SF can be called "nature," in Aristotle's understanding, the first as potential and the second as actual, and the actual nature of a thing is what specifies it as a natural kind and further stabilizes it in its being. Thus the SF not only informs PM and makes of it a complete substance, but it also makes the substance be what it is, that is, it organizes and specifies it; again, it confers on the protomatter a stable mode of being, so that the natural substance underlies its accidents in more than transitory fashion.

The accidents of a natural body are shown in Figure 1, arranged, somewhat arbitrarily, around the inner core. They are grouped into

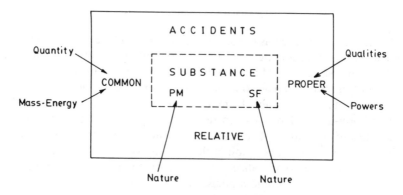

Figure 1. A NATURAL BODY

three categories: common, proper, and relative. The most important of the common accidents is quantity, shown next to protomatter, and the most important of the proper accidents is quality, shown next to the specifying form. The relative accidents then include relation and the last six species of accident listed in Aristotle's *Categories*, all of which are of lesser importance for our purposes.[13] With regard to quantity, I would merely point out that for St. Thomas, it, along with protomatter, is the individuating principle in a natural substance: *materia signata quantitate*. I wish to go beyond him and suggest that modern science has furnished us with a measure for this *quantitas materiae*, namely, mass-energy, to which we shall return later.[14] Furthermore, with regard to quality as a proper accident, this stands for distinctive attributes through which we come to know a nature. A thing's actions and reactions enable us to ascertain its powers, and from these we judge its nature, or say what it is.

Let me elaborate on this last statement, which is crucial for this essay, by proposing a series of diagrams that schematize the relationships between characteristic activities, natural powers, and the stable natures that are their source. I shall consider stable natures as pertaining to three broad genera: inorganic natures, plant natures, and animal natures; in addition I shall discuss human nature as adding a further specification to animal nature. These are stable in the sense that substances with these natures have a fairly permanent mode of being; they are to be distinguished from transient natures, about which more will be said below. In each case the diagrams are based on the information contained in Figure 1, but further elaborated to show the differences in the various types.

Figure 2 is the schematic model for an inorganic nature.[15] Here the protomatter (PM) is shown in the center, and the specifying form (SF_i) as a series of concentric circles radiating outward from it. The impression should be that of an organizing field structuring the matter and endowing it with particular qualities or powers. Among inorganic natures I would include those of elements, compounds, and minerals. Unlike medieval and Renaissance Aristotelians, however, I would not regard elements as perfectly homogeneous, endowed only with active or passive qualities such as hot and cold, wet and dry. Instead I would attribute to them and their compounds the four basic forces or potentials of modern physics, and see each of these as natural powers from which emanate characteristic activities and reactivities. Thus the electromagnetic force (EF) functions in chemical reactions, the gravitational force (GF) in gravitational interactions, the weak force (WF) in radioactive emission, and the strong force (SF) in nuclear interactions. In light of these capabilities, I would be disposed to see the specifying form (SF_i) as an efficient cause in at least some of these activities, operating through its powers in much the way the substantial form of a living organism causes its life activities.

A plant nature differs from an inorganic nature precisely in its life functions, and these are diagrammed in Figure 3. The medievals thought such functions were produced by three vegetative powers, the nutritive, the augmentative, and the reproductive; these are conserved in the model but presented in a slightly different way. Instead of three powers we assign four, shown directly above the four powers of the inorganic. These are: homeostatic control (HC), which maintains the plant in equilibrium with its environment; metabolic control (MC), which regulates its processes of food and energy conversion; developmental power (DP), which controls cell differentiation and growth; and reproductive power (RP), which enables the mature organism to form new members of the species. Note that all of the powers of the nonliving are conserved in the plant organism, only now they are unified under a distinctive specifying form, SF_p, otherwise known as the plant soul. This form is again modeled here as an organizing field, structuring the matter and providing it with powers that can animate its component parts. In this way it accounts for both inorganic and organic activities; it

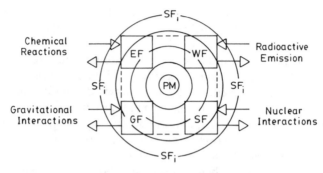

SF: STRONG FORCE
EF: ELECTROMAGNETIC FORCE
WF: WEAK FORCE
GF: GRAVITATIONAL FORCE

Figure 2. AN INORGANIC NATURE

controls all of the plant's physical and chemical components, for example, and enlists them in the service of its health and survival.

Animal natures have yet more capabilities than plant natures, and these are modeled in Figure 4. The distinguishing powers are those of sensation and emotion, which here are shown at the top of the diagram, directly above the vegetative powers and those of the inorganic. The two topmost are the external senses (ES) and the internal senses (IS), which receive stimuli from the outside world and convert them into sense knowledge; the two below these elicit an emotional or behavioral response (BR) to the knowledge received, and so stimulate the motor powers (MP) to the appropriate movement. As heretofore, the specifying form, SF_a, called the animal soul and again shown as the field correlate of protomatter, organizes and stabilizes all of the powers – sensitive, vegetative, and physico-chemical – to assure the well-being of the organism.

Human nature is an animal nature, and thus it is modeled in much the same way, as shown in Figure 5. The only difference is the addition of intellect (I) and will (W) to the top two lines of the diagram, the first enabling the human being to attain abstract thought and to reason, the second to make free decisions and implement them in a responsible way. These two powers are most distinctive in that they are immaterial

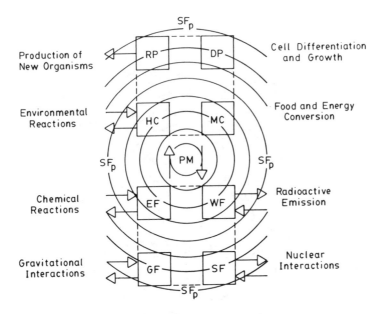

Figure 3. A PLANT NATURE

and spiritual in their being and operation. All other powers we have analyzed thus far function through the use of chemicals or organ systems. Neither intellect nor will, on the other hand, has any organ through which it operates. They are completely "psychic," that is to say, they are not body-soul powers, but are uniquely powers of the human soul, the highest type of substantial form (SF_h). We have again modeled this form as an organizing field, energizing not only intellect and will but all the lower powers as well. This is the soul that Catholic teaching maintains is directly created by God, as will be seen shortly when we discuss the problems of creation and human generation.

These, then, are the main types of stable natures in the world of nature, each respectively signalled by the presence within it of an inorganic form (SF_i), a plant form (SF_p), an animal form (SF_a), or a human form (SF_h) – the last three also known as a plant soul, an animal soul, and a human soul respectively. Were a person presented with a natural body and asked what nature it possessed, his answer would have to be based on how the body acted or reacted in a variety of circum-

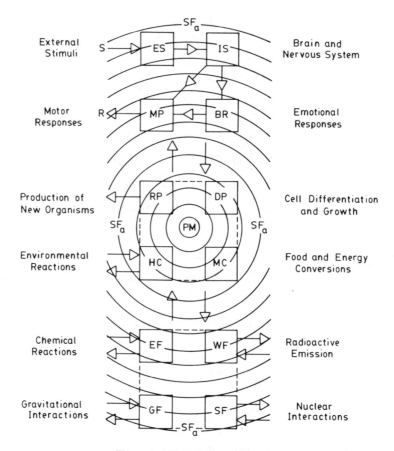

Figure 4. AN ANIMAL NATURE

stances. These activities or reactivities would sooner or later reveal to him the powers from which they proceed, and on the basis of these powers he would have to judge whether the body had an inorganic, plant, animal, or human nature. *Agere sequitur esse*: action follows being, or, a thing acts as it is. This, unfortunately, is the only method we have of ascertaining natures. I am speaking now of stable, mature natures, not those of developing organisms, but promise to return to that topic shortly.

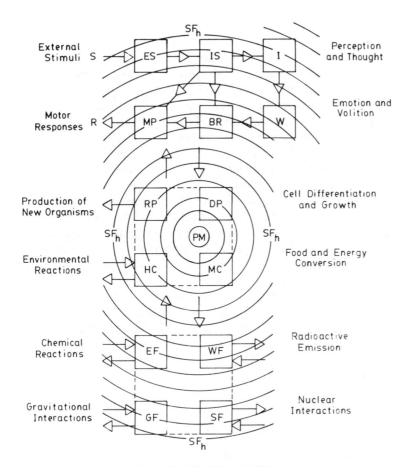

Figure 5. HUMAN NATURE

As opposed to stable natures there is also the problem of transient natures, the type of being the medievals referred to as an *ens viale*, a being "on the way," as it were. They did not have extensive knowledge of such entities, using the expression to refer mainly to seeds, *semina*, which quite obviously were on the way to becoming plants or animals. Their interest was mainly in problems of classification, and they solved that neatly enough by classifying seeds in the same category as the plant or animal it would normally produce. Now, among the great discoveries of modern science, in my estimation, is its uncovering an astounding number of transient entities in the physical universe. I refer to the world of elementary particles, most of which have a very transitory existence. Practically all of them are radioactive, with an extremely short half-life; those that are not are usually charged particles, whose very charge renders their independent existence precarious, to say the least. A realist philosopher of science thinks of these as more than *entia rationis*, as having some mode of existence outside the mind. Do they also have natures? Can we speak of the nature of a proton, a neutron, an electron, or a quark, or of the many types of hadrons and leptons that have been identified in high-energy physics? On the basis of my model for inorganic natures I would tend to answer in the affirmative; the four basic forces are operative in the subatomic realm as well as in the atomic and the molecular, and the same store of mass-energy is available for their formation as for the higher types. But immediately I would qualify my answer and say that these are not stable natures, like those of elements and compounds, but rather transient forms that emerge from the potency of protomatter under more or less violent conditions and then recede back into that potency, only to be replaced by other emergent forms.[16] Their organizing form is not a stabilizing form like those of the natures we have been discussing, and yet for the brief period of its existence it is a specifying form (SF_t). That is why I refer to them as transient natures, enjoying in most cases but a fleeting existence, but still a part of the world of nature.

A CREATION-EVOLUTION SCHEMATIC

These prenotes will have to suffice for our examination of material-formal causality; now we turn to the more difficult problems of efficient-final causality to reopen the question of how nature operates as an agent cause. A convenient way to begin it to revise the account of beginnings already given by incorporating into it the findings of modern science relating to the origin of the universe and the theory of evolution. Again I shall make use of a diagram, shown in Figure 6 and labeled "A Creation-Evolution Schematic."

In this diagram I sketch a concordist view of Catholic theology and modern science, associating God's creative act at the beginning of time with the "big-bang" theory of cosmic beginnings. Time t_0 began some ten to twenty billion years ago with the production by God, *ex nihilo*, of

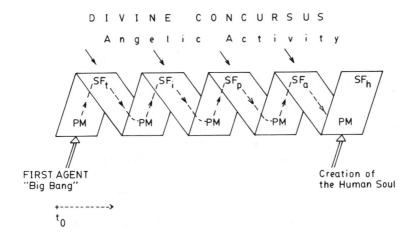

Figure 6. A CREATION-EVOLUTION SCHEMATIC

the primordial mass-energy of which the universe is now composed. Perhaps simultaneously with the act of creation, God as First Agent or Prime Mover also initiated the "big-bang," releasing the enormous energy of the primitive mass for the formation of the natures now found in the universe. This was a cosmic explosion, and if we accept recent astrophysics, its vestiges are still discernible at the edge of our expanding universe.

Before entering into the subject of cosmic evolution, we may inquire into the efficient cause of this expansion. My option is that the same agent as initiated the explosion also continues to propel its radiation to the present day. How it does so is difficult to explain, but I have no qualms about accepting *impetus* or some other inertial principle as the instrumental cause of the expansion. Medieval and Renaissance thinkers had reservations about this, but after the mechanics of Galileo, Newton, and Einstein, I do not see how we can have such reservations in the present day. Should we take the First Agent to be God, then of course we have the divine *concursus* to support the entire universe in its being throughout all of time, and that should dispel any ontological worries we may have about its continual expansion.

Cosmic evolution is slightly more difficult to explain. In Figure 6 I give the elements of a generic picture by indicating in succession along the time-axis the five types of substantial form already discussed. These are, in order, transient natures (SF_t), inorganic natures (SF_i), plant natures (SF_p), animal natures (SF_a), and human nature (SF_h). They correspond to the stages of evolution commonly accepted among scientists: the period of fundamental particles impelled at high energy; that of element and compound formation; the two periods of biogenesis, wherein first plants and then animals were generated; and finally that of hominization, wherein *homo sapiens* first appeared. Note that, apart from the divine *concursus*, there is no direct intervention by God in the universe until the moment of man's appearance, for it is then that the creation, *ex nihilo*, of the human soul (SF_h) takes place – an act that, according to Catholic teaching, will be repeated as long as the human race continues to be procreated.

For each substantial form its companion principle, protomatter (PM), is also indicated in Figure 6, thus preserving the essential composition of natural substances diagrammed in Figure 1. Leaving aside, for the moment, the human soul (SF_h), on the assumption that it is produced directly by God, we make ask where these natural forms come from. The answer supplied by Thomistic philosophy is a surprising one: they are not preexistent as forms, nor are they created in any way; instead, they are simply "educed" from the potency of protomatter. This is the famous doctrine of the *eductio formae ex potentia materiae*, which holds that all natural forms (man's excluded) are already precontained in the potentialities of the substrate, and require only the action of the appropriate agent to bring them forth into being. The analogy of the sculptor may enable us to understand this. We may ask where the form of David existed before it was chiseled out of marble by Michelangelo. One answer would focus on the exemplary cause: the form existed in the mind of the sculptor. But an equally valid answer would look to the material cause, to the block of marble, and say that David's form was resting in there all along, simply waiting to be led forth, educed, liberated from the matter under the action of Michelangelo's chisel. In a proportionate way, natural forms may be said to be resident in protomatter, their exemplars already present in the divine mind, awaiting only the proper agent to confer on them actual existence. And if we propose mass-energy as the metric of protomatter, corresponding to the medievals' *quantitas materiae*, we already have a measure in terms of which we can quantify many aspects of the subsequent development.

We may now run quickly through the various stages of evolution shown in Figure 6. At the initial creation, God created the transient natures we know as elementary particles; these we can still "recreate" in the present day by educing them from protomatter with our high-energy particle accelerators. Such natures existed for the briefest period of time, but already elaborate theories have been formulated to explain how vast numbers of nuclides (1700 are known at present) were formed from them in the first minute, say, of the universe's existence. When the proper conditions of radiation and energy obtained, the chemical elements were formed from these in the relative abundances they manifest to the present day: hydrogen in greatest abundance (75% of the mass of the universe); helium next (most of the remaining percentage); then the other elements (contributing only slightly to the total mass). With that, stable inorganic forms (SF_i) had appeared and the first stage of evolution was complete. From this flux were formed the stars and the planets some five billion years ago, and on some planets, at least on Earth, molecules we now classify as "organic" were being formed. Significantly, no special efficient causality was required for this initial development, all of it taking place under the impetus of the primordial "big bang."

The next stage, that of biogenesis, saw living forms, first plant (SF_p) and then animal (SF_a), make their appearance. Since these are "material forms," in the terminology of the scholastics, they likewise are educed from the potency of matter. It is for this reason that a single causal line is shown in Figure 6, first educing inorganic forms from protomatter, then educing plant forms from composites of SF_i and PM, and finally educing animal forms from composites of SF_p and PM. One may wonder whether all of this could come about by chance, by random collision, as it were, or whether some guiding intelligence was necessary to channel this line of efficient-material causality so as to form such complex organisms. It is here that St. Thomas's *quinta via* becomes applicable: the intelligence of the Author of Nature, who programmed protomatter in the way he did, is certainly an indispensable component of the evolutionary process. But I have also indicated along the top of Figure 6, under the divine *concursus*, a place for angelic activity acting throughout time. This is to accommodate Aquinas's view that separated substances, the angels we know through Holy Scripture, take part in the governance of the universe (*Summa Theologiae*, la:qq.110-114). This would surely be an elegant role for them, bringing the lower natural forms to the summit of hominoid perfection, although from a metaphysical standpoint they are not absolutely necessary, since the First Cause can do alone anything that can be effected through secondary causes.

The final stage of cosmic evolution, then, is hominization, the appearance of man with his immaterial soul (SF_h). Here there is a break in the single line of causation, because, as immaterial, the human soul cannot be educed from the potency of matter. The entire process of evolution, as diagrammed thus far, can bring organisms to a level just below that of thought and volition, but they cannot progress to the final stage. Here God's creative act is required. So we indicate this with a second input of divine causality, the production, *ex nihilo*, of the human soul, tailored to match the ultimate disposition of matter as this has been prepared, over billions of years, for its reception.

The foregoing account, it should be stressed, gives only a generic picture of the evolution of natures, grouping all species of elements and compounds under inorganic natures (SF_i), all kinds of vegetative life under plant natures (SF_p), and all types of sensitive life under animal natures (SF_a). We must now take up the problem of individuation-speciation and consider how individuals of various species or natural kinds in each of the broad kingdoms (mineral, plant, and animal) come into existence. This I shall refer to as natural generation. For this there is much more information to work with, since it is the ongoing process whereby individuals are added to each species, but only a small amount of this can be considered in what follows.

NATURAL GENERATION

Assuming that the basic concepts of material-formal causality are already at hand, the main problem continues to be that of tracing the lines of efficient-final causality through which a new nature is produced. The simplest cases are the types of synthesis and decomposition studied in inorganic chemistry, whereby, for example, a compound is produced through the combination of elements or is decomposed into them. This type of process can be schematized in the same way as the evolutionary process in Figure 6, and thus recourse will continue to be made to that model. Instead of redrawing it each time, however, since the PM-component remains the same and all forms (the human excepted) are educed from the potency of PM, it will only be necessary to indicate the succession of forms involved in a particular generation. The convention for doing so will be to present the line of forms in sequence, as shown in the following rewriting of the content of Figure 6:

$$\Rightarrow SF_t \rightarrow SF_i \rightarrow SF_p \rightarrow SF_a \Rightarrow SF_h. \qquad [1]$$

Here the arrows with double shafts represent God's creative action (first at the beginning of time and then at the instant of hominization), whereas those with single shafts represent the eduction of a form from the potency of PM, in the eduction process already explained.

The generation of a compound from elements can now be schematized, following this convention, by dividing the inorganic form (SF_i) into two types, one for elements (SF_e), the other for compounds (SF_c). Then the formation of a compound from two elements, say, may be represented as follows:

$$SF_{e1} + SF_{e2} \rightarrow SF_c. \qquad [2]$$

If $e1$ is hydrogen and $e2$ is oxygen, then the compound produced is water; if $e1$ and $e2$ are sodium and chlorine the compound is salt, and so on. All of these reagents have different natures, as manifested by their different properties, and when the new nature is generated, in the same process the old nature ceases to be. Thus water is neither hydrogen nor oxygen, though it was formed from them, nor is salt sodium or chlorine, though it was formed from them also. One may now ask: what is the agent or the efficient cause that produces water from hydrogen and oxygen or salt from sodium and chlorine? My answer would be that the essential cause of the generation of a compound is its elemental components, which do so in virtue of the powers with which their natures are endowed. These are shown generically in Figure 2: the main factors are electromagnetic force (EF) and gravitational force (GF), the first more than the second, acting through the electronic outer shells of the elements. There are also accidental causes, such as the agents that bring the elements into proximity and remove the impediments to their interaction. (Note the similarity here to the two types of causality invoked by Aristotle to explain gravitation as a natural motion: a *per se* cause, the generator of the nature; and a *per accidens* cause, the *removens prohibens* that enables the nature to exercise its powers and initiate the motion [*Physics*, VIII:4].)

Let us now look further into the problem of how individual natures, as opposed to specific natures such as water and salt, are produced. We can do so by considering the similar mechanism involved in the decomposition of water, through electrolysis, into hydrogen and oxygen. In this case the process can be symbolized by the reverse of [2],

$$SF_c \rightarrow SF_{e1} + SF_{e2},\eqno[2']$$

which would be written by the chemist, in his notation:

$$2\,H_2O \rightarrow 2\,H_2 + O_2.\eqno[2'']$$

This is usually understood to mean that two molecules of water break down, under electrolytic action, into two diatomic molecules of hydrogen and one diatomic molecule of oxygen. The bearing of this on the question of "individual natures" is seen when we consider the two hydrogen molecules that result from the breakdown: each has the same nature, and yet one is distinct from the other. Whence does this differentiation arise? Assuming, from the brief discussion of Figure 1 above, that the principle of individuation is matter signed with quantity, one can say that the extrinsic agent so alters the quantitative dispositions of the water molecules that it is impossible for these to break down, for example, into one molecule each of hydrogen and oxygen. The matter of H_2O is so "signed" by its quantity that the only way mass-energy requirements can be satisfied is by the eduction from the protomatter of two hydrogen natures, each with a different but equal mass-energy, along with one oxygen nature. In biological terminology we might say that the hydrogen molecules are twins, identical twins, but chemists are not concerned about this; for their science it is sufficient to consider the nature apart from the individual, since in their view all hydrogen molecules are necessarily the same.

Now let us turn to a more complicated case, one closer to the concerns of this group, which is taken from nuclear chemistry: the natural radioactive breakdown of one element to another. We may start with the uranium series, wherein the element uranium ($_{92}U^{238}$) breaks down through the series of elements: thorium ($_{90}Th^{234}$), protactinium ($_{91}Pa^{234}$), radium ($_{88}Ra^{226}$), etc., to lead ($_{82}Pb^{206}$). This process may be diagrammed as follows:

$$SF_{e1} \rightarrow SF_{e2} \rightarrow SF_{e3} \rightarrow SF_{e4} \rightarrow SF_{e5},\eqno[3]$$

wherein the numerical indices refer to the five elements in the order given. This would seem to be a better instance of natural agency than the decompositon of water, for it is the natural powers of these elements, and particularly the weak force (WF) and the strong force (SF) plus the others shown in Figure 2, that serves to explain the ongoing radioactivity. No other agents would seem to be required, apart from the First Agent that initiated the process of element-formation, as shown in Figure 6. (On the other hand, were we to consider the neptunium series, which starts with $_{93}Np^{237}$, since this element is produced artificially in a laboratory, the series itself may be regarded as artificial, being initiated by whatever incidental agent functioned in the eduction of neptunium from the potentiality of protomatter.)

If radioactive breakdown is truly an instance of natural agency, we may further inquire how quantitative dispositions function in the eduction of the new natures that are produced. Here the answer is similar to that already proposed for chemical decomposition. Because of a basic instability in the heaviest nucleus in the series, in this case changes in both nuclear structure and outer electron shells are involved. These must always satisfy the quantitative dispositions (essentially mass-energy requirements) of the substrate, with the result that the natures of thorium, protactinium, radium, and ultimately lead, are successively educed from within its potentiality.

Another interesting problem arises here: that concerning the difference between stable natures and transient natures mentioned earlier in this essay. In [3] we have shown uranium, thorium, etc., as stable elemental forms and not as transient forms. Hence the question: Should they not be indicated as transient, since their natures are unstable? If the answer is affirmative, the process should have been diagrammed:

$$SF_{t1} \rightarrow SF_{t2} \rightarrow SF_{t3} \rightarrow SF_{t4} \rightarrow SF_e, \qquad [3']$$

where the first transient elemental nature, SF_{t1}, gives way to the second transient elemental nature, SF_{t2}, which gives rise to the third, and then the fourth, until we come to the fifth, which is not a transient elemental nature but rather that of a stable element, SF_e. Although both the "stable nature" and the "transient nature" explanations would seem possible, I prefer the stable [3] over the transient [3'], for the following reason. Despite their nuclear instability, all of these elements have chemical natures and properties that are identifiable and stable, and are so listed in the Periodic Table. (A similar problem obtains with isotopes of other elements: most of these have radioactive nuclei, but still they are regarded chemically as endowed with the same nature as the element in its stable form.) Thus I favor the "stable nature" option, although the "transient nature" option supplies us with a mechanism that will have a further application, as we shall see presently.

This brings us finally to biological generation, where transient forms again present a problem, and where the nucleus, not of the atom but of the cell, figures importantly in its solution. The main difference between biological generation and the other generations already discussed is that the former clearly involves seeds, the *entia vialia* of the scholastics, and thus transient natures now must enter explicitly into the discussion.

To begin, let us propose the following simple model of plant generation:

$$SF_{p1} \rightarrow SF_{tp} \rightarrow SF_{p0}. \qquad\qquad [4]$$

Here $p1$ refers to the parent plant, tp refers to the transient plant-like nature of the seed throughout its development, and $p0$ refers to the offspring, numerically different from $p1$ but pertaining to the same species. The agency involved in the first transition, from SF_{p1} to SF_{tp}, is a natural agency associated with the powers of plant life as shown in Figure 3. Chemical materials are absorbed by the parent organism, $p1$, through homeostasis (HC) and metabolism (MC), and then with the aid of its developmental (DP) and reproductive power (RP), to form the genetic materials contained in the seed. These processes are carried out under the aegis of the parent nature; after that, upon separation from the parent, a form of life persists in the seed while it begins its own internal development and effects the transition to the offspring, $p0$.[17] This is done through a transient plant-like form, SF_{tp}, that draws nourishment to the incipient organism and directs its growth, resulting in the gradual formation of the organ systems that are the correlates of the powers found in the species. These represent quantitative dispositions analogous to the nuclear and electronic structures of chemical natures. When these dispositions are suitable for sustaining a stable, individual plant nature, that is, when material defects have been eliminated and twinning or recombination, say, can no longer occur, the new plant form SF_{p0} endowed with its own proper powers is educed from the potentiality of the protomatter. The entire process involves only plant natures, either stable or transient, both pertaining to the same species throughout.

More complexity is introduced when we consider a model for the generation of a higher animal, for here the transient nature may have to undergo more than one stage of development in order to produce an organism of the particular species. Let us assume at the outset two cases. The first is a one-stage development, analogous to that of the plant, which may be shown as:

$$SF_{a1,2} \rightarrow, SF_{ta} \rightarrow SF_{a0}, \quad\quad\quad [5]$$

where $a1,2$ refers to the stable animal natures of the parent organisms, ta to the transient animal-like nature of the seed (thus similar to [4]), and $a0$ to the animal nature of the offspring. The second case would be a two-stage development, which may be written as:

$$SF_{a1,2} \rightarrow SF_{tp} \rightarrow SF_{ta} \rightarrow SF_{a0}, \quad\quad\quad [5']$$

where $a1,2$ and $a0$ are as in [5], but tp stands for a transient plant-like nature and ta for a transient animal-like nature, intermediates between the two. It is noteworthy that St. Thomas, without knowing any details of modern science, favored [5'] over [5], thus indicating that this option is not incompatible with the principles of his natural philosophy (*Summa Theologiae*, la:qq 90 & 118). Attention has been directed to his teaching in the present day because of its association with theories of delayed hominization, to be discussed in the following section.

To explain the process detailed in [5], one could say, on the analogy of plant generation, that the agency involved in the production of SF_{ta} is a natural agency originating within the animal powers shown in Figure 4. Chemicals are absorbed as food, synthesized through the metabolic power (MC), and encoded in egg and sperm by the reproductive powers of the parents (RP). The preparatory stage in the formation of the transient nature occurs when sperm and egg cease to be informed by the parents' natures and take on life-like characteristics of their own. Under natural agency or through other means, depending on the manner of insemination, the sperm fertilizes the egg and produces the zygote, which is endowed with the animal-like nature of its particular species, ta. Then, through the incipient powers of that animal-like nature, the zygote differentiates and grows until, again by natural agency, the stable nature of the animal appears, at which point the offspring, $a0$, has been produced. This is a long and involved process, going well beyond the period when twinning or recombination can occur, and involving the formation of organ systems for homeostasis (HC), metabolism (MC), and developmental power (DP), as well as those for sensation (ES and IS), appetition (BR), and movement (MP), which involve an incipient nervous system and possibly limb articulation. Finally, when the proper quantitative dispositions have been assured, the stable animal nature is educed from the potency of the substrate, and a new individual results.

The alternative process shown in [5'] involves two transient forms, one associated with the production of a plant-like nature, SF_{tp} , which, at the first stages of the animal's existence, exercises only the three basic powers of plant life (HC, MC, and DP, in Figure 4). As cells divide and the organism builds up more complex systems, however, this first transient nature is replaced by a second, SF_{ta} , which is able to exercise the basic powers of animal life (self-movement, MP; sensation, ES and IS; and behavioral reaction, BR). Under its guidance the organism continues to develop and grow until the proper quantitative dispositions are reached, at which point the stable animal nature is educed from the potency of matter and a new individual of the species, $a0$, results. Note that the same line of agent causality is behind the process of generation in [5'] and [5], the only difference being in the types of transient natures involved, which is mainly a matter of formal causality. There is a difference that bears on efficient causality, however, for the earlier activities of the organism suggest only the presence of a plant-like form as its organizing and conserving principle, whereas its later activities suggest the presence of an incipient animal form as their proportionate cause.

In view of the fact that the foregoing models, from [4] to [5'], indicate only generic plant and animal natures and do not take speciation into account, we must emphasize that in each case we have taken the transient form to pertain to the same species as that of the parent organism. Thus we are not adopting here an evolutionary perspective, but rather one that respects natural history as we presently experience it, where oaks come from other oaks, chimpanzees from other chimpanzees, and so on. Within our perspective the question that should be addressed is how a new, individual, stable nature of the particular species comes into being, not how something other than an oak or a chimpanzee can come from the genetic materials of those species. The precise point is that a new individual chimpanzee might *not* be produced in the order of nature, but rather that material defects might lead to the abortion of the incipient nature, or that something might happen within the developmental process that would lead to two chimpanzees being formed rather than one – the problem of twinning, which takes on special importance in the case of human generation.

Returning for a moment to the generative process shown in [5'], note the resemblance between it and that for radioactive decay shown in [3] and [3'], particularly in [3'], where several transient natures are involved. In both cases a problem arises concerning the intermediate forms. Is the transient form that is becoming a chimpanzee really a chimpanzee from the first moment of formation of the zygote? Alternatively, is the radioactive neptunium that is breaking down into lead really lead from the first moment of its radioactive decay? Invoking the *agere sequitur esse* axiom, one would think that, just as an element with the properties of neptunium should be regarded as neptunium and not as lead, even though it will eventually become lead, so a transient nature that exhibits only vegetative activities should be regarded as a plant and not as an animal, even though the term of its growth is a stable animal nature.

NATURAL NORMS FOR MEDICAL ETHICS

With this, all of the materials are in place to raise the question of natural norms for medical ethics. In discussing the case of animal generation we have already adumbrated that of human generation, and thus we return to the problem of hominization, not in the context of creation but in that of procreation. Our concern is similar to that already diagrammed in Figure 6, for now we are concerned with the famous question: does ontogeny recapitulate phylogeny, i.e., does the individual human arise from nature in much the same way as the human species took its origin? To make the problem more explicit, we shall model human generation in the two ways we have presented animal generation:

$$\mathrm{SF}_{h1,2} \to \mathrm{SF}_{th} \Rightarrow \mathrm{SF}_{h0} \tag{6}$$

and

$$\mathrm{SF}_{h1,2} \to \mathrm{SF}_{tp} \to \mathrm{SF}_{ta} \Rightarrow \mathrm{SF}_{ho}. \tag{6'}$$

Here the main difference is that the subscript a has been replaced by the subscript h. The peculiar difficulty for medical ethics respects that substitution, however, for SF_h stands for the human soul, which cannot be educed from the potency of matter but must be created directly by God. This requirement arises from the powers proper to human nature, shown in Figure 5, which requires more than those for vegetative and sensitive life, but intellect and will as well. As immaterial, these powers and the nature endowed with them must have an immaterial source. Moreover, granted God's necessary activity, the question arises as to when this takes place. At what moment in time does God create the human soul and infuse it into the developing organism? The process shown in [6] would imply that this moment occurs very early in the generative process, possibly as soon as conception has taken place; this is usually known as immediate hominization. That shown in [6'], on the other hand, implies that God's activity takes place only after a period of evolution (analogous to the schema in Figure 6), and this is known as delayed hominization.

There is not space in this essay, nor have we the competence, to list and evaluate the pro's and con's of these alternatives. Suffice it to mention that the main arguments for immediate hominization are that all the genetic materials from which a new human person will be formed are already present in the fertilized egg, and that the only efficient causality to effect a new human being is the sperm's meeting the egg, which terminates when fertilization has occurred. The arguments for delayed hominization, on the other hand, are based on research in reproductive biology, mainly on studies of cell division, twinning and recombination, implantation, and so on. St. Thomas's authority is usually invoked in its support, although it is admitted that the explanations of medieval biology have practically nothing in common with those of the present day.

The conclusion suggested by this study is that a Thomistic natural philosophy, updated to incorporate the findings of modern science, can influence the decision between the two alternatives, and would probably favor delayed over immediate hominization on the basis that the former is more consonant with nature's other operations. The main argument involves a renewed understanding of the causality of nature, particularly in the interplay between powers and natures and the ways in which material and efficient factors interact in natural processes. The fact that all the genetic materials are present in the zygote, for example, is not decisive: in chemical and nuclear processes all of the materials are usually present from the beginning, without their necessarily influencing the terminal outcome. Similarly, in natural generation, the factors that bring the gametes together can be merely incidental causes; they need not enter essentially into the formation of a new organism, any more than the chemist need enter into the formation of a new compound.

The peculiar problem presented by hominization really relates to the divine economy, namely, whether God would produce the summit of his creation – a new person, an immortal human soul – without having the proper quantitative dispositions present to match it and stabilize it in being. Throughout the entire course of nature new substances are educed from the potency of protomatter only when such conditions are met. Nature does not operate with big jumps, but induces its changes only gradually. Indeed, if the theory of evolution is correct, most of the history of our universe was spent building up such dispositions to prepare for the advent of the highest form that can be received into bodily existence. There is no theological reason to hold that the human soul comes to be individuated in other than a natural way: matter signed with quantity is the basic principle, and one human twin is differentiated from the other in much the same way as are hydrogen molecules. In this view of things, God creates the individual immaterial form with a transcendental order to quantitative dispositions already present in an incipient human form. The moment he does so a new person has been created and a new individual added to the human species.

It may be argued that the foregoing accent on genetic materials or quantitative dispositions is not sufficient to account for human generation, that a proper efficient agent is necessary to explain the on-going unity of operation within the developing organism. This agent would be a differentiated central organ that is already teleologically programmed to produce a unique human being, and on this ground could be regarded as a goal-oriented efficient cause of future embryological development. The argument would seem to be strengthened by the fact that the historical and functional identity of this central organ can be traced continuously from the zygotic nucleus to the organizing center of the plastula and the gastrula, then to the primitive streak, then to the central nervous system, and finally to the brain. On this view, as soon as fertilization is complete, the nucleus of the zygote would have begun to control all subsequent development as an efficient cause, and for this reason the zygote itself should be regarded as human, that is, as already informed by a human soul. This would require God's creation of the soul and its infusion into the zygote upon the completion of fertilization, before the developmental process has itself begun, which is effectively the same as "immediate" hominization.

The line of argument developed earlier in this paper is that no special efficient causality of this type, apart from the causality of the parent organisms, is necessary to explain the early or even the later development of the incipient organism. The search for a proper efficient cause for organic "buildup" in this case runs into the same difficulty as the search for a proper efficient cause for the "breakdown" of a heavy element through natural radioactivity. It misconstrues how nature itself operates as a cause in the eduction of forms, and would bring God into a natural process at a stage where his action is not required. As long as the genetic materials are themselves human, that is, are not programmed to produce a chimpanzee, they have within themselves all that is required to develop a new individual of the human kind. Defects might be encountered, however, and natural powers might be thwarted in their efforts to prepare the materials properly for the infusion of a human soul. Alternatively, what started out as a single mass might end up as supplying the proper dispositions for the generation of more than one individual, in which case more than one human soul would be created by God and infused into different organisms at the term of the generative process.[18]

The account here offered, it must be stressed, does not intend to legitimate induced abortion through the use of external agents acting on genetic materials, any more than it would legitimate contraceptive measures that might prevent egg and sperm from meeting in the natural reproductive act. No one currently holds that human egg and human sperm are animated by a human soul, and yet there is no denying that they are human gametes and should be treated accordingly. Similarly, even though the human zygote may be informed by a transient animating principle, it is still informed by a plant-like or animal-like nature that is programmatically human and so pertains reductively to the human species. This is the sense in which it is an *ens viale*, admittedly at a more advanced stage than egg or sperm, but still "on the way" to becoming a fully human organism. As reductively human it must be regarded with even more care than the genetic materials from which it was formed.

The foregoing line of reasoning, to be sure, does not pretend to be apodictic. Apart from general principles and analogies based on better known aspects of nature's operations, there is little hard data on which to base the theory of delayed hominization. Here natural philosophy has to await further studies in reproductive biology, for only these will cast light on the problem of transient natures as they may be found in human generation.

It should follow as a corollary that those who attempt to formulate legal or moral norms relating to human reproduction are limited in what they can say with any degree of practical certitude. The U.S. Supreme Court legalized abortion during the first trimester of pregnancy at least partially on the basis that there was no consensus among philosophers on when human life definitely begins. Medical ethicists labor under a similar handicap. The Catholic magisterium favors immediate hominization, but I doubt that this is because theologians in Rome have better knowledge of natural philosophy or of reproductive biology than do scholars in research centers throughout the world. My view is that Catholic teaching on the time of hominization and associated issues is more disciplinary than veridical – establishing an *orthopraxis* that is on the "safe side," as it were, safeguarding human life when in doubt, rather than risk its irresponsible destruction.

Are there, then natural norms for medical ethics? In principle, yes, but considerably more work has to be done in both natural philosophy and biological science before they can be seen as safe and providing a *recta ratio* for responsible practice.

The Catholic University of America
Washington, D.C., U.S.A.

NOTES

[1] This is explained in my *The Role of Demonstration in Moral Theology: A Study of Methodology in St. Thomas Aquinas*[14].

[2] See the essay, 'A Thomistic Philosophy of Nature', in my *From a Realist Point of View: Essays in the Philosophy of Science*[8].

[3] See 'Defining the Philosophy of Science,[8],pp. 1–21.

[4] I have published materials in support of this thesis in a number of works, among which the more significant are *Prelude to Galileo: Essays on Medieval and Sixteenth – Century Sources of Galileo's Thought*[13]; *Galileo and His Sources: The Heritage of the Collegio Romano in Galileo's Science*[9]; 'Three Classics of Science',*The Great Ideas Today 1974*,[15],pp. 211–272. See also the chapter I have written entitled "Traditional Natural Philosophy" in *The Cambridge History of Renaissance Philosophy*, forthcoming.

[5] My own version of this is set forth in a mimeographed text, *Praelectiones de philosophia naturali seu cosmologia*[12].

[6] The grounds for such a philosophy of science are established in my *Causality and Scientific Explanation*, 2 vols., Ann Arbor: University of Michigan Press, 1972–1974, reprinted Lanham, MD.: University Press of America, 1981, but again out of print.

[7] The more important of these studies are 'The Intelligibility of Nature: A Neo – Aristotelian View'[10] and 'Nature as Animating: The Soul in the Human Sciences'[11]; see also 'Galileo and the Causality of Nature' in [13],pp. 286–299.

[8] Full details are given in [6].

[9] In this meaning of soul, which is envisioned by Aristotle in his *De anima*, Book III, there are plant souls and animal souls as well as those in human beings.

[10] On this principle, set out in *Physics* VII, 1 and VIII, 4, see the essays by J. A. Weisheipl in [18], pp. 75–120.

[11] E.g., in his commentary on Aristotle's *Physics* and elsewhere.

[12] See the essays cited in note 8, as well as two additional papers presented at the Woodrow Wilson International Center for Scholars, Washington, D.C., in 1984, referenced in 'Nature as Animating', [11], p. 612, n.1.

[13] *Categories*, cc. 7–9; see B. M. Ashley, [5], pp. 297–307, for the relevance of the Aristotelian categories to the subject of this paper.

[14] On *quantitas materiae*, see J. A. Weisheipl, 'The Concept of Matter in Fourteenth Century Science',[17], pp.319–341; on measures of mass, etc., see my essay "Measuring and Defining Sensible Qualities', in [8], pp. 73–97.

[15] Fuller explanations of these models are given in 'Nature as Animating'[11] and in the papers cited in its first footnote.

[16] Some discussion of this will be found in my essay, 'Elementarity and Reality in Particle Physics',in [8], pp. 185–212.

[17] Aquinas referred to this as an "active force" it receives from the parent, giving it a plant nature "in first act,"*Summa Theologiae*, 1a, q. 118, a.1 and ad 4um. In what follows we refer to this incipient form as a "plant – like" nature, since it lacks the full perfection of a plant form.

[18] This may provide the answer to the objections raised by B. M. Ashley in his 'A Critique of the Theory of Delayed Hominization',[4], pp. 113–133. I wish to thank Fr. Ashley for his careful reading of a draft of this essay and for a number of suggestions that led to its improvement.

BIBLIOGRAPHY

1. Aquinas, St. Thomas: 1964, *Summa Theologiae*, Blackfriars, McGraw-Hill, New York.
2. Aristotle: 1931, *De Anima*, trans. J. A. Smith, Clarendon Press, Oxford.
3. Aristotle: 1955, *Physics*, trans. W. D. Ross, Clarendon Press, Oxford.
4. Ashley, B. M.: 1976, 'A Critique of the Theory of Delayed Hominization', in D. G. McCarthy and A. S. Moraczewski (eds.), *An Ethical Evaluation of Fetal Experimentation: An Interdisciplinary Study,* Pope John XXIII Center, St. Louis, pp. 113–133.
5. Ashley, B. M.: 1985, *Theologies of the Body: Humanist and Christian*, Pope John XXIII Center, St. Louis.
6. Landen, L.: 1985,*Thomas Aquinas and the Dynamism of Natural Substances*, unpublished Ph.D. disssertation, The Catholic University of America, Washington, D.C.
7. Wallace, W. A.: 1981, *Causality and Scientific Explanation*, 2 vols., University Press of America, Lanham, MD.
8. Wallace, W. A.: 1983, *From a Realist Point of View: Essays in the Philosophy of Science*, 2nd ed., University Press of America, Lanham, MD.
9. Wallace, W. A.: 1984, *Galileo and His Sources: The Heritage of the Collegio Romano in Galileo's Science*, Princeton University Press, Princeton, N.J.

10. Wallace, W. A.: 1984, 'The Intelligibility of Nature: A Neo-Aristotelian View', *Review of Metaphysics* **38**, 33–56.
11. Wallace, W. A.: 1985, 'Nature as Animating: The Soul in the Human Sciences, *The Thomist* **49**, 612–648.
12. Wallace, W. A.: 1960, *Praelectiones de philosophia naturali seu cosmologia*, Dominican House of Philosophy, Dover, MA.
13. Wallace. W. A.: 1981, *Prelude to Galileo: Essays on Medieval and Sixteenth – Century Sources of Galileo's Thought*, Kluwer Academic Publishers, Dordrecht, Holland.
14. Wallace, W. A.: 1962, *The Role of Demonstration in Moral Theology: A Study of Methodology in St. Thomas Aquinas*, The Thomist Press, Washington, D.C.
15. Wallace, W. A.: 1974, 'Three Classics of Science', in *The Great Ideas Today, 1974*, Encyclopaedia Britannica, Chicago, pp. 211–272.
16. Weisheipl, J. A.: 1980, 'The Axiom *Opus naturae est opus intelligentiae* and its Origins', in G. Meyer and A. Zimmerman (eds.), *Albertus Magnus Doctor Universalis: 1280–1980*, Matthias Gruenwald, Mainz, pp. 441–463.
17. Weisheipl, J. A.: 1963, 'The Concept of Matter in Fourteenth Century Science', in E. McMullin (ed.), *The Concept of Matter*, Notre Dame University Press, Notre Dame, pp. 143–175.
18. Weisheipl, J. A.: 1985, *Nature and Motion in the Middle Ages*, ed. W. E. Carroll, The Catholic University of America Press, Washington, D.C., pp. 75– 120.

DAVID C. THOMASMA

THE HUMAN PERSON AND PHILOSOPHY OF MEDICINE:
A RESPONSE TO WILLIAM A. WALLACE

Fr. Wallace has come out with both guns blazing. He has an extraordinary capacity for sketching the big picture. He did not fail us here. Setting
out for himself the task of integrating ancient, medieval, and modern
science as the basis for a new philosophy of human nature, he would
undergird a modern Catholic medical morality. At one point in his essay
he cautions that "one must know the history of science down to the
present, and be able to evaluate it critically, to judge what has been
transient and ephemeral and what will retain lasting importance for
future ages" ([11], p. 25). I can think of no one better able to match
these abilities in Catholic circles than Fr. Wallace. But he wisely
cautions: "And the person who embarks on this enterprise must be
prepared to be contradicted at every turn along the way" ([11], p. 25). This
is, indeed, the case in any interdisciplinary and systematic effort like his.

Insight and courage are required to embark on this task. Fr. Wallace's
project displays both. His insight is that a *via media* can be constructed
between pluralism and relativism on the one hand and genetic deter-
minism of sociobiology (as articulated by Edmund Wilson) on the other.
This *via media* would base ethics on a philosophy of human nature so
that metaethically judged good and evil would be established by actions
in accord with and opposed to one's own nature.

He notes elsewhere:

The regulative idea is simple: nature is an intrinsic principle of perfective activity, and the
better we understand a nature or a natural kind the more we can appreciate how it should
act. Thus we would bridge the "is" and the "ought" by rooting the norm for action in an
objective standard: a nature that is not completely refractory to understanding ([12],
p. 613).

Reconstruction on a new basis is needed for medical ethics because a
plurality of principles alone is insufficient. Nonetheless, the philosophy
of human nature as passed on in Catholic teaching is woefully inadequate because it is based on a discredited late medieval science and
worldview. Consequently, the great interdisciplinary project for

55

Edmund D. Pellegrino et al. (eds.), Catholic Perspectives on Medical Morals, pp. 55–60.
© *1989 Kluwer Academic Publishers. Printed in the Netherlands.*

Catholic moral thinking is to rebuild a valid philosophy of human nature on our newest conceptions and models of science.

Following Dr. Wallace's approach, the rudderless moral ship of our post-Enlightenment pluralistic age, perhaps best described by Alasdair MacIntyre in *After Virtue*[4], would again be set on course. MacIntyre claims that without moral consensus in principle, or the moral authority of the Church, Western civilization is condemned to increasingly shrill assertion and counterassertion.

Of course, many other twentieth-century thinkers have constructed attempts to base an ethics on a new anthropology. John Dewey, Max Scheler, Sartre, and the Structuralists, although quite disparate thinkers, immediately come to mind. The results of these efforts are not exactly what Fr. Wallace has in mind, of course, What their authors had hoped would encapsulate the heart of a moral life has instead contributed to the very pluralism of our times. Their constructs become additional targets at the skeet-shoot of life. With the possible exception of Dewey, who built his ethic on natural science and psychology, the efforts of most other thinkers have had a very short half-life in the United States. In view of so many conflicting claims for allegiance, it is no wonder that Ortega y Gasset suggested that reading the history of philosophy was like a trip to a pleasant insane asylum[6].

Dr. Wallace's effort is different from other efforts on several accounts. And this is where his courage is shown. In an age that not only mistrusts natural science, and that daily adds to its cubit an anxiety about the impact of medical technology on persons, he boldly anchors morality in a scientific synthesis. It is not that he does not recognize the crisis of science. Elsewhere he has written:

One might characterize the late twentieth century as a period when men have become oblivious of nature. Not only is the concept of human nature under attack, but the broader awareness of nature itself . . . is no longer part of our mental equipment. The ecological crisis and the near exhaustion of many natural resources bear eloquent witness to this state of affairs. The scientific and industrial revolutions have made us proficient at converting the objects that surround us into artifacts . . . but they have dulled our appreciation for the intelligibility of nature in its own right ([10], p. 36).

Among the key points in the synthesis Dr. Wallace constructs are the claims that there are both stable and transient natures ([11], pp. 30–36), that motion and change occur by reason of an internal force through the specifying form, even in inorganic matter ([11], pp. 31–32), and that evolution can provide for a creationist intervention at the Big Bang and then at the inauguration of each human soul ([11], p. 37, p. 45). Nature is known to the extent that we can model it ([12], p. 616), and extensive models are certainly provided. As his thesis unfolds, it becomes clearer and clearer that Fr. Wallace is a kind of Aristotelian-Thomistic Teilhard!

At the heart of his thesis is the role of transient nature in the development of organisms – the transient nature provides growth and development until "the stable animal nature is educed from the potency of matter and a new individual of the species . . . results ([11], p. 46). This model of generation is then applied to delayed hominization. In this process the intellect and will, the human soul, is unable to be educed from matter. Fr. Wallace's argument for delayed hominization, that nature does not operate from "big jumps, but induces its changes only gradually" ([11]), p. 49), is firmly based on current reproductive biology, and even other natural changes in inorganic matter he describes. He suggests that Catholic teaching about the moment when human life begins is an "orthopraxis" rather than one based firmly on scientific knowledge. I agree. He also hints that medical ethical discourse should be based on knowledge of the realities about which the discussion and pronouncements pertain. Further, if this knowledge is absent, no pronouncements should be made.

Given the "indulgences" with which Fr. Wallace works, e.g., that causality occurs in nature, and the realist assumption that sub-atomic particles exist in nature, his account is remarkably successful, brilliant, clever, and challenging. But there are other "indulgences" or assumptions in the account that are not spelled out.

Let me explicate these as matters of concern. First is that Catholic moral teaching has been and should be based on the natural law. Because of the myriad objections to natural law theory, indeed, as Fr. Wallace himself knows, objections to a realistic model of the whole scientific enterprise, is the natural law assumption a wise one at all? In particular, what can the sciences of human behavior, missing in Fr. Wallace's schematic (but noted by him elsewhere) ([12], p. 631, 642), contribute to judging right from wrong on the basis of our nature? If anything, civilized conduct seems to be in a constant evolutionary flux, much like the generation of new forms in nature. For this reason, church pronouncements and medical ethics opinions based on natural law theory seem more and more hopelessly out-of-date. One is left trying to defend an archaic philosophy as well as the teaching itself. This is

especially difficult to do when the audience, many of them fellow
Catholics, does not accept the tenets of that philosophy.

Second, the creationist assumption is everywhere present, especially as
regards the origin of the universe and the human soul. Thus, it is better
to describe Fr. Wallace's "natural" philosophy as a Christian, indeed,
Catholic philosophy of nature – as he intended, a philosophy in service
of a theological vision. The assertion about the psychic, non–physical
powers of the intellect and will are, in this context, assumed to be non–
controversial, although in an apologetic context it would be ([11],p. 33).

Lurking, too, in the creationist assumption is the idea of a person-
centered universe. We might call this homocentrism, or personacentr-
ism to avoid sexist connotations. Note how God's creative act is described, a production *ex nihilo*, "tailored to match the ultimate
disposition of matter as this has been prepared, over billions of years,
for its reception" ([11], p. 40). If this faith vision is accurate, then the
dying cancer patient's body is as beautiful as the firmament, the processes in it as violent as the vast incomprehensible explosions of Cigna, the
person whose body it is, the apogee of all creation. Yet what can this
vision tell us about decisions to be made at the bedside in accord with
this nature? Should we withdraw food and water because the will of the
patient would be honored[3], or should we continue because this personal life-form, now in a coma, deserves such care and respect[5]?

Although Fr. Wallace applied his synthesis to the problem of hominization, and thereby to the question of abortion, many other steps are
needed to link the philosophy of human nature to morality, as he noted
at the end of his essay. I would like to see expanded the generative,
developmental insights nature conveys rather than the static forms, a
particular strength of Fr. Wallace's account. This would have a far-
reaching impact on standard natural law ethics, for it could incorporate
growth and development over time in ethical analysis, something sadly
missing in the more static natural law theory.

Outside of the Catholic tradition, something he did not intend, would
Fr. Wallace's project be successful? As much as I would hope so, my
concern is that it would not be. And this is based on two major grounds.

First, the creationist assumptions would appear to be precisely a *deus
ex machina*. Why do we need creation to explain generation of the soul,
for example? And how are the powers of specifying forms, even transient forms, different in living matter from genetic codes? Although I do
not think all social behavior can be traced to these codes, much of what

passes for natural law can be traced to animal behavior and genetics[1].

Further, in an evolutionary perspective such as that advanced by Teilhard de Chardin, Fr. Wallace must account for the appearance of new forms and of new *levels* of forms, inanimate, animate, vegetative, animal, and human. While it is true that nature only proceeds gradually, there is a qualitative leap between inanimate and animate matter and between vegetative and animal. Although Fr. Wallace acknowledges the difference, this level-jump is not discussed. If new levels can be educed from lower level interactions (perhaps through a bundle of lesser causes), then why cannot we assume, as Teilhard did, that the pyschic, non-physical dimension of human life is also present to be educed from lower matter itself? Can life emerge from the primordial oceanic soup? Does matter contain non-material dimensions? Even if it did, or if the soul were instead created, what possible moral difference would it make? If the intellect and will are so independent of matter, could they not be instilled at any time? Would our stand against abortion be enhanced or diminished if, irrespective of whether the fetus were a person, its living organization, its transient form were seen as already harboring within it our psychic powers much the way Jung's cultural unconsciousness was seen as harboring the powers of good and evil[2]?

The second ground for concern about Fr. Wallace's effort lies in the nature of the profession of healing and the nature of illness[8]. An ethic of a profession, the medical morality addressed at this conference, must take into account not only the nature of its subject matter – human persons who need healing – but also the interactions of values occurring at the basis of the doctor-patient relation, the family, institutions, society, and religion. This interaction requires greater attention to the values we construct from the protomatter of our lives, their constant adjustment in concert with life-challenges and others we meet, their hierarchical and axiologial order[7].

Historically speaking, because natural law theory concentrated on the "big picture", it tended to neglect this interpersonal dimension altogether. But this is to pick on only one theory. In fact, can it be the case that all philosophical medical ethics is inadequate because it neglects the graced dimension a cry for healing creates? In my view this can be seen as almost a sacramental reality. Is not this movement also part of the openness of our nature to the transcendent?

If the developmental aspects, the transient natures, of Fr. Wallace's accouunt are more vigorously pursued as I have suggested, we may have a golden opportunity to combine our best scientific and philosophical thinking with our scriptural heritage, for the Kingdom of God proclaimed to be at hand is a community of growth and healing, not just a collection of stably-formed human persons. The crux of Fr. Wallace's account here is that nature exercises its internal powers, divinely granted and divinely aimed (also [12], p. 615). Perhaps the spiritual possibilities of life and matter are yet to be extruded from the evolutionary process. The past and present God may also be, as Schillebeeckx has asserted, the eschatological future of human beings[9].

Loyola University Medical Center
Maywood, Illinois
U.S.A.

BIBLIOGRAPHY

1. Ellos, William, S.J.: 1986, 'Sociobiology and Structuralist Ethics', in J. Wind and V. Reynolds (eds.), *Essays in Human Sociobiology*, V.U.B. Study Series No. 26, Brussels, Vol. 2, pp. 9–16.
2. Jung, C.J.: 1970, *Psychological Reflections*, Jolande Jacobi (ed.), Bolligen Series 21, Princeton University Press, Princeton.
3. Lynne, Joanne (ed.): 1986, *By No Extraordinary Means*, Indiana University Press, Bloomington and Indianapolis.
4. MacIntyre, A.: 1981, *After Virtue*, Notre Dame University Press, Notre Dame, Ind.
5. Meeting of the Pontifical Academy of Sciences: 1985, 'Ethical, Medical and Legal Questions on the Artificial Prolongation of Life', Declaration adopted by scientists, *L'Observatore Romano*, Nov. 11, p. 10.
6. Ortegay Gasset, J.: 1960, *What is Philosophy?*, Norton, New York.
7. Pellegrino, E. D., and Thomasma, D.C.: 1988, *For the Patient's Good: The Restoration of Beneficence in Medicine*, Oxford University Press, New York.
8. Pellegrino, E. D., and Thomasma, D.C: 1981, *A Philosophical Basis of Medical Practice*, Oxford University Press, New York.
9. Schillebeeckx, E.: 1969, *God, the Future of Man*, Sheed and Ward, New York.
10. Wallace, William A.: 1984, 'The Intelligibility of Nature: A Neo-Aristotelian View', *Review of Metaphysics* **38**, 33–56.
11. Wallace, William A.: 1989, 'Nature and Human Nature as the Norm in Medical Ethics', in this volume, pp. 23–53.
12. Wallace, William A.: 1985, 'Nature as Animating: The Soul in the Human Sciences', *The Thomist* **49**, 612–648.

BRUNO SCHÜLLER

PHILOSOPHICAL FOUNDATIONS OF CATHOLIC MEDICAL MORALS

Biomedical ethics constitutes one of the many kinds of applied normative ethics, as do economic ethics, political ethics, or peace ethics. In
discrimination between such kinds of ethics, one can hardly proceed
from the rules of a logical division (*divisio logica*), and so it is to be
expected from the start that they will partially overlap. Political ethics
may have among its topics segments of biomedical ethics, possibly under
the rubric "health politics". There can, however, be hardly any doubt
about where the philosophical foundation of biomedical ethics is to be
found: namely, in the foundation of every kind of applied ethics, that is,
in general normative ethics. What T. Beauchamp and J. Childress
discuss in their book, *Principles of Biomedical Ethics*[3], for example, is
chiefly general normative ethics in the tradition of W. Frankena[9] or R.
Brandt[4].

For this reason the topic of my essay places me in a predicament. How
could I in the space of an essay set forth the basic concepts of normative
ethics? I would have to appeal to not very helpful generalities, and apart
from that I would simply have to repeat thoughts I have already
published elsewhere. Allow me, therefore, instead, to confine myself to
relating some observations I have made about the study of normative
ethics and about the discussions and controversies carried on within it.
How are these controversies conducted? Why do they so often fail?
Granted, what I am about to say is indeed not new. But as John
Passmore writes: "Philosophers, as it happens, need a lot of reminding"
([21], p. 8).

I. DEONTOLOGICAL OR TELEOLOGICAL NORMATIVE ETHICS?

Of what importance is it for biomedical ethics whether moral judgments
are proffered teleologically or non-teleologically, that is, deontologically? If I wanted to express myself in the manner of speech most
familiar to me personally, then my answer to this question would be: It
is of great importance if one thinks *deontologically* in the manner of an

Edmund D. Pellegrino et al. (eds.), Catholic Perspectives on Medical Morals, pp. 61–78.
© *1989 Kluwer Academic Publishers. Printed in the Netherlands.*

Immanuel Kant or of J. G. Fichte; but if one argues deontologically in the manner of W. D. Ross, William Frankena, or John Rawls, then it might be of little importance, if not completely immaterial.

Have I expressed myself in the question and answer in such a way that I can reasonably expect it to be understood, at least by ethicists? Hardly. Certainly the word pair "teleological-deontological", insofar as it serves as the designation of ethical theories, has been employed by philosophers for decades. But whoever pays careful attention to how this term is used by this or that philosopher soon discovers that it is often used equivocally. Thus one finds confirmed G. E. Moore's remark in his *Ethics* about the use of the term "utilitarianism":

> One of the difficulties which occur in ethical discussions is that no simple name which has ever been proposed as the name of an ethical theory, has any absolutely fixed significance. On the contrary, every name may be, and often is, used as a name for several different theories, which may differ from one another in very important respects. Hence, whenever anybody uses such a name; you can never trust to the name alone, but must always look carefully to see exactly what he means by it ([18], p. 30).

An example: as everybody knows, John Rawls explicitly contrasts his *Theory of Justice*[23] with utilitarianism. In a critical assessment of this theory, Brian Barry[2] calls attention to the fact that it essentially agrees with the theory put forward by R. M. Hare in *Freedom and Reason*[10]. Hare himself has confirmed this. But he counts himself among the utilitarians. And Ian Narveson had already, in 1967, equivalently said that Rawls' "Justice as Fairness" is nothing other than rightly understood utilitarianism[19]. I myself share this opinion and number Rawls among my allies, inasmuch as I also plead for a utilitarian theory, though I regularly call it, of course, *teleological* and indeed for the same reasons that moved Friedrich Paulsen in his time to replace "utilitarian" by "teleological". The adjective "utilitarian" has on the Continent, as a rule, a strongly deprecating meaning, even among philosophers.

Therefore, even philosophical terms such as "utilitarian" and "deontological" are used equivocally. What happens when one does not take notice of that? In Germany, Robert Spaemann has offered a destructive criticism of *teleological ethics* as it has been advocated for some time by many Catholic moral theologians. In a footnote he explicitly cites F. Bockle, F. Scholz, and me. How does he proceed in his criticism? He replaces the term "teleological ethics", since it is not unequivocal, with the ostensibly unequivocal term "utilitarianism" and then refutes utilitarianism as *such* by adopting the very criticism which J. Rawls, M. Singer, and D. H. Hodgson, among others, have proposed ([23, 25, 11]). Suppose this criticism is conclusive. Must we moral theologicans look on ourselves as refuted in the matter? Not in the least. For none of us is committed to what Spaemann calls utilitarianism. It is remarkable that Spaemann does not take account of the equivocal nature of "utilitarianism". In G. H. von Wright one reads the phrase: "Neither Hume, or the British utilitarians, nor Moore" [31]. Here Moore is not considered one of the utilitarians. Why not? Does von Wright not regard him as British? That is rather improbable, since von Wright occupied Moore's professorial chair at Cambridge. J. Plamenatz, in *English Utilitarians*, writes: "Though Professor Moore places to his own credit some unnecessary dialectical victories, most of the arguments in *Principia Ethica* are, I think, convincing. And if that is so, then utilitarianism, however qualified to make it look more plausible, is an untenable theory" ([22], p. 147). Moore, otherwise a utilitarian par excellence, is not counted as a utilitarian but rather as a victorious champion against utilitarianism. Apparently, neither Plamenatz (All Souls College, Oxford) nor von Wright has accepted the term "Ideal-Utilitarianism", although H. Rashdall (Oxford) coined it in 1907 for the designation of his own theory, a theory such as G. E. Moore also puts forward, J. N. Findlay regards the term coined by Rashdall as excellent and notes that for a long time "Ideal-Utilitarianism" has often been used to designate the ethical thinking of Plato and Aristotle.

In 1971 Richard Norman published *Reasons for Actions: A Critique of Utilitarian Rationality*. This book was especially stimulating for me, not least because, guided by the Table of Contents, I began not with the first page but with page 93, the second chapter, "Integrity and Utility: The Religious Ethic and the Humanistic Ethic", which opens thus: "The utilitarian can give no distinctive sense to such concepts as 'conscience,' 'guilt,' and 'remorse.' He may perhaps employ these concepts, but he will have to define them simply as heavily charged versions of 'regret.' For the account which he gives of moral praise and blame is in terms of their utility: when other people do wrong we blame them to make them better"[20]. That is the classical argument against those who deny human beings the power of free self-determination. From controversial theology it is familiar as criticism of Catholic theology on Luther's doctrine of "servum arbitrium". R. M. Hare's prescriptivism might also be affected by it. Now Hare considers himself a utilitarian. But does

Luther also count as a utilitarian? Hardly. Judging from this argument, what can Norman understand by "utilitarian"? In his second chapter, Norman treats the attitude of "reverence" towards the "sacred". There he explains that: "'Sacred' is a value-word, and to call something 'sacred' is to ascribe a certain kind of value to it. But a non-teleological value. If one regards x as sacred, this does not mean that it is valued as an end, for this is not the kind of value . . . that can be produced or increased at all" ([10], p. 101). The expression "non-teleological value" somewhat astonishes me. For according to the Aristotelian-scholastic axiom, "quod habet rationem boni, habet rationem finis," the terms "teleological ethics", "axiological ethics", and "agathology" are often used simply as synonyms. But it is completely clear what he understands by "end" and "teleological". He connects with it only that meaning which R. S. Downie and E. Telfer also suggest when they write of Kant's second form of the categorical imperative: "There is something odd about speaking of *persons* as ends"[6]. However, apart from that, what can Norman understand by 'utilitarian' when he asserts that a utilitarian cannot substantiate the morally required attitude towards the 'non-teleological values'? Must a Christian utilitarian admit that he can give no reason why he should love God (the non-teleological value *per eminentiam*) with his whole heart and soul? He who now opens the book to the first page finds that there "is the theory that all reasons for performing actions must ultimately be derivable from statements of human wants or desires or satisfactions. I shall refer to it as the 'utilitarian' view of practical rationality, thereby using the term 'utilitarian' in a somewhat wider sense than is usual"[20]. To me, this interpretation of the adjective "utilitarian" is so strange that I would have to study the entire book in order to understand the explanation. This is additional evidence for the homonymy of the word "utilitarian".

Whoever, like Kant and the authentic magisterium of the Catholic Church, maintains that the use of artificial contraceptives and also homologous artificial insemination are morally illicit (on the basis of their own descriptive properties) seems to show thereby that he thinks deontologically. But does it follow therefrom that anyone who maintains that these ways of acting are morally indifferent, can only do so insofar as he adheres in general to a teleological ethics? Hardly. But it seems to me that, after all, many Catholic philosophers think in this way and believe that when they oppose teleological ethics, they affirm at the same time that the use of artificial contraceptives has to be considered morally wrong. But how do they attempt to refute teleological ethics? Above all, by counter-example: citing the alleged killing of an innocent person as apparently permitted or required from a teleological viewpoint.

Recall that in the year 1810 the Royal Dutch Society of Haarlem posed a prize question: "Why do philosophers widely differ on first principles but agree on the conclusions and responsibilities they derive from their principles?" John Stuart Mill has openly stated that "People who agree about specific moral judgments frequently disagree about the first principles of morals" ([17], p. 35). John Finnis, a determined deontologist, and I, a determined teleologist, are both of the opinion that the sheriff of a county must not execute a provably innocent scapegoat, even if only through that means could he save the lives of other innocent people. I argue that this is illicit according to a kind of *lex lata ad praecavendum periculum generale*, which for teleological reasons allows no exceptions. J. de Lugo and other theologians justify in this way the exceptionless obligation of the confessional secret. Finnis thinks nothing of this argument. I reject his criticism of this argument. Therefore, we agree about a specific moral judgment even though we disagree about the first principles of morals. Still, suppose it were the case that I must concede that I was not successful in providing a conclusive teleological reason for my opinion. It would not occur to me to give up a theory for which I think I have good reasons merely because I see myself unable to refute a counter-example. Why should that which I cannot refute *de facto* be irrefutable in principle? The reasons I have for teleological ethics appear to me to have such weight that in a gnoseological sense they justify the presumption that such a counter-example can be removed in principle.

How far does it hold for deontologists like Frankena and teleologists or utilitarians that they disagree about the first principles of morals? In a chapter entitled "The Ethics of Love" Frankena quotes the theologian Garnett, who makes the following point against Emil Brunner: "Justice is built into the law of love, since in its second clause, it requires us to love our neighbors as ourselves or equally with ourselves." Concerning this Frankena remarks: "If we so construe the law of love, it is really a twofold principle, telling us to be benevolent to all and to be so equally in all cases. Then, the ethics of love is not purely agapistic and is identical with the view I have been proposing" ([9], p. 58). Frankena accordingly numbers himself among the deontologists because he sees in the commandment of charity two combined ethical principles. Let us compare this with John Stuart Mill's remark in *Utilitarianism* ([17], ch. 2): "As between his own happiness and that of others, utilitarianism requires him [the agent] to be strictly impartial as a disinterested and benevolent spectator. In the golden rule of Jesus of Nazareth, we read the complete spirit of the ethics of utility. To do as you would be done by, and to love your neighbor as yourself, constitute the ideal perfection of utilitarian morality." In what way does the utilitarian, Mill, differ from deontologist Frankena? Perhaps Mill only considers the Golden Rule as a single principle. R. S. Downie and E. Telfer in *Respect for Persons* write: "Such an attitude[i.e., *agape*] is morally basic not only in that it is paramount, but also in that all other moral attitudes and principles can be explained in terms of it" ([6], p. 37). In what do they differ from Mill? Quite certainly they would refuse to be placed among the utilitarians. I, for my part, share their view and for exactly that reason consider myself a teleologist or "ideal" utilitarian.

Despite the significance that may belong to the division of ethical theories into teleological and non-teleological theories, it does seem to me often more important to examine the various forms of argumentation for their internal soundness. Whoever thinks teleologically, as I do, cannot deduce from this that he is therefore immune from all errors in reasoning, which, since the time of Aristotle, have been identified and dealt with repeatedly. It is obviously no objection, then, to a deontological theory if one of their resolute advocates is guilty in this or that case of an *ignoratio elenchi* or a *petitio principii*.

II. PARENESIS AND NORMATIVE ETHICS

For many years, I have been under the impression that numerous discussions and controversies among many ethicists are not distinguished by their logical rigor but rather by a surprising absence of logic. How is that to be explained? At the beginning of the opening chapter of his *Ethics*, Frankena calls Socrates "the patron saint of moral philosophy" ([9], p. 2). Does he wish to say thereby that moral philosophers should emulate Socrates as their model and consider as their task what Socrates viewed as his vocation? If so, then I must state that Frankena himself (so far as I am familiar with his writings) does indeed follow Socrates as portrayed in Plato's *Crito*. In this dialogue he deals with normative ethics. But judging by Plato's *Apologia*, Socrates' life-task concerned "religious welfare" or the "the care of souls". In *Apologia* Socrates says of himself:

> I will never stop philosophizing and exhorting you and making things clear to everyone whom I meet, while in my customary way I speak: "My good man, you are an Athenian . . . and you are not ashamed to care for your property and its continual increase . . . yet for the knowledge of that which is good and true and for your soul that it become as good as possible do you not care and are you unconcerned about it?" And when one of you disputes that . . . I will . . . question, examine, and refute him; and if he seems to me not to progress toward *Arete*, but only to claim he does, I will chide him that he esteems the trivial rather than that which is the most valuable . . . I will deal thus with young and old, whomsoever I meet . . . . For you know that my God commands me so; and I think that there has never been in our city a greater good for you than service to my God (29d–30a).

As W. Jaeger remarks on this text, Plato here conducts the philosophizing of Socrates, insofar as it is care of souls, back to two main forms of speech, to the *elenchus* and the *protrepticus*, both linked to the oldest form of the *parenesis*[1] ([12], p. 601). [If one recalls the sermon of John the Baptist as it is summarized by the synoptics, then one can easily recognize a prophetic variant of this characteristic sequence of the *parenesis*: *Elenchos* = Brood of vipers! The axe is laid to the roots of the tree. *Protreptikos* = be repentant, produce the appropriate fruits.]

I cannot recall that Frankena's *Ethics* reflects the parenetic way of a Socrates or even of a John the Baptist. Disciplined in his reasoning and extremely careful in his judgment, he pursues normative ethics and metaethics. Why does he set himself this limit in his *imitatio Socratis*? Because he is an academic teacher, a professor of philosophy? Is the task of the care of souls thereby incumbent on him? I assume that Frankena does not think so. But other academic teachers of ethics also follow Socrates in that they hold forth strongly in *parenesis*. Whether and to what extent they do that rightly I cannot discuss further here. It is, after all, quite simple to keep them at bay whenever one has a mind to do so. For that, one only needs to opt for obduracy. However, the situation changes once an ethicist takes his *parenesis* for an argument within normative ethics. But this happens all too often. The result then consists in false reasoning such as *petitio principii*, circular argument, and the like.

For a long time, the decalogue has been for me a paradigm of parenetic discourse. How are its commands understood by someone who is well-versed in the task of normative ethics, but for some reason has not yet perceived the peculiarity of parenetic language? Some months ago I found two excellent illustrations of this. Bertrand Russell remarks on the fifth commandment:

> "Thou shalt do no murder" would be an important precept if it were interpreted as Tolstoy interprets it, to mean "thou shalt not take human life." But it is not so interpreted; on the contrary, some taking of human life is called "justifiable homicide." Thus murder comes to mean "unjustifiable homicide;" and it is a mere tautology to say "thou shalt do no unjustifiable homicide." That this should be announced from Sinai would be as fruitless as Hamlet's report of the ghost's message: "There's ne'er a villain, dwelling in all Denmark, But he's an arrant knave" (I, 5, 124).

J. D. Mabbott expresses himself similarly:

> In the authorized version of the English Bible (1611) the ten commandments in Exodus XX include, "Thou shalt not kill." Now here there is nothing in the word "kill" to make this tautologically self-evident, like "moneys owed ought to be paid." And, of course, the difficulty arose that the Old Testament is full of killings of which God is held to show no disapproval. So in the Revised Version (1885) the commandment reads: "Thou shalt do no murder." But now what does "murder" mean? Does it not mean those killings which are reprehensible, which ought not to be committed? . . . "Murder" has a morally condemnatory sense and therefore the Revised Version commandment is a tautology ([15], p. 33).

Russell and Mabbott read the decalogue as the answer to the question, What should we do?, the basic question of normative ethics. If it were meant as the answer to this question, then actually it turns into a series of tautologies. In reality it must be read as divine *parenesis*. Ephesians (4: 25 ff) says: "So from now on, there must be no more lies: you must speak the truth to one another. . . . Even if you are angry, you must not sin. . . . Anyone who was a thief must stop stealing!" Here we are not dealing with a sequence of tautologies but rather with a sequence of admonitions, which, of course, within normative ethics would be tautologies. And these tautologies are probably often overlooked, since they have a special sense when they are taken as warnings. Another reason that tautologies are often not recognized in ethics discourse is that people frequently do not seem to grasp the *evaluative* meaning of certain terms and turns of phrase. Again, Mabbott asserts: "It is clear, however, that some rules are not tautologies – 'Tell the truth,' 'Honor thy father and mother'" ([15], p. 33). Granted, "Tell the truth" may not be a tautology in certain contexts. Yet in the passage cited from Ephesians, "speak the truth to one another" obviously establishes the contrary of "telling lies"; is it not then a characteristic of that virtue that it presents the opposite of the vice of lying? The expression "to be quite honest about it" is as a rule synonymous with "to speak the truth openly". But 'honesty' is the name of a virtue. "Honour thy father and mother," at least insofar as it is a rendering of the fourth commandment, is as tautological as "Thou shalt do no murder," when it is understood or – more correctly – misunderstood as the answer to the question: "What should we do?" For "father" and "mother" are in this case prescriptive names, so that the commandment must be read as in Romans (13: 7): "Render to all their dues . . . honour to whom honour is due!"

It seems important that we pay attention to the synonyms of the most general moral predicates of value — "good and bad", "right and wrong". Previously, such synonyms were "natural and unnatural", "reasonable and unreasonable". For some time now we have often used the terms "responsible and irresponsible", "humane and inhumane". It may have to do with the use of the last – mentioned word pair that the substantive "dignity of man" or "dignity of a person" can even replace the term "ethics".

U. Eibach has written a book with the title, *Medicine and the Dignity of Man*. This is qualified by the subtitle, *Ethical Problems in Medicine*. Next year, the German-speaking moral theologians will be convening to confer on the topic: "Migration and the Dignity of Man". That is to be read as "Ethical Problems of Migration". In a paper on genetic research a moral theologian characterizes the task of the ethicist as follows: "to critically attend to the technological development process and to make himself permanently the advocate of people, their dignity, their rights but also their duties" ([7], p. 437). The sentence is somewhat pleonastic and must be subdivided in this way: "Make himself advocate of the people, that is, of their dignity, that is, of their rights and duties." It is clear, in order to explain the task of the ethicist that one has to define "ethics", but defined and defining must be synonymous. However, is the ethicist the *advocate* of the dignity of man, his rights and duties? Is he that in his capacity as ethicist? The moral theologian cited above seems to regard himself as the conscience of the genetic researchers, and he addresses a protreptic discourse to them, probably without being explicitly aware of it. Thus, for example, he puts forward four moral principles that should be observed in genetic research: (1) the principle of universality – the positively stated Golden Rule; (2) the principle of fairness, summarized as follows by John Rawls: "Under like circumstances everyone must be treated equally, or one must at least be ready oneself to take over the role of each individual interested in the entire action" ([7], p. 438); (3) the principle of human dignity; (4) the principle of responsibility. I fear these are not four moral principles but rather four linguistically different expressions which signal one and the same principle – Kant's categorical imperative – which defines the moral point of view. Therefore, the ethicist is supposed to remind the genetic researchers of what it means to behave ethically not only in all spheres of life, but also in the domain of genetic research. He enjoins them: "Even as genetic researchers you have to act impartially, justly, fairly, and with conscious responsibility, and to respect the dignity of humans." Who would have thought that?

What is achieved by reference to human dignity in the field of normative ethics? The importance of human dignity for normative ethics is reflected in a report on in-vitro fertilization that appeared in 1984. "No human being is a property." What that means has become clear, historically, with the abolition of slavery. "Historically, slavery was a social institutionalization regarding certain human beings as properties of others. A slave was his master's property . . . he was instrumental, and so could be used, exploited, or even killed." With the abolition of slavery, people came to recognize the fundamental equality of all human beings and with it also the principle of "respect for persons". This respect is a fundamental requirement of justice, in virtue of which no human being is to be used or exploited for any purpose whatsoever. It recognizes that every human being has at least the right not to be used merely as a means to the needs or interests of others, and that every innocent human being has at least the right not to be killed. But perhaps the utilitarian tradition does not think so. For it, "respect for the individual human being" is not the paramount moral value; the paramount moral value is, rather, "the greater good for the greatest number". This allows that, given certain circumstances, the life or lives of individual human beings may be regarded as instrumental, and so expendable for that "greater good". Ultimately, for utilitarianism, all values can be "traded off"; human beings themselves may be expendable.

Whoever can think and write in that way is hardly "sicklied o'er with the pale cast of thought" (*Hamlet*, III, 1, 85). Is the institution of slavery actually suitable for interpreting the sentence, "No human being is a property"? Is that not reminiscent of the well-known fallacy, "obscurum per obscurius explicare"? Many facts indicate that the institution of slavery is first of all to be understood from the standpoint of ethical particularism. But ethical particularism recognizes the principle of "respect for persons". Its advocate simply has not yet conceived that every human being, just because he is a human being, is a person to whom one owes respect.

He knows that one "should love his neighbor as himself." He differs from the ethical universalist only by his answer to the question: "Who is my neighbor?" If the *New Testament* evidently speaks universalistically, just as, for that matter, do the Stoics, and nevertheless approves of the institution of slavery, then it thus is clear that the expression "slave" is used equivocally. Thus, for example, it is used in one sense by Aristotle in his *Politics* and in a different sense by the author of the Letter to the Colossians. When Paul writes (4, 1): "Masters, make sure that your slaves are given what is just and fair, knowing that you too have a Master in heaven," then he uses "slave" in a legal sense and not in the sense of the given thesis that a slave can (may) be exploited or even killed.

Consequently, there are historical views of slavery that appear consistent with the principle of "respect for persons". Or must we assume that even Paul, insofar as he does not reject every form of slavery, is not aware of or does not appreciate this ethical principle?

But suppose, *dato, non concesso*, that the acknowledgement of the principle "respect for persons" is equivalent to the negation of all slavery, how must one then think of utilitarianism when according to it "human beings themselves may be traded off"? Is it not thereby entailed by utilitarianism that human beings are fundamentally only slaves, and indeed slaves for the greatest number? I doubt whether it is fair to characterize utilitarianism this way when for quite a long time every decent human being has deeply abhorred all that slavery means.

The sentence, "No human being is a property," modifies only Kant's second formulation of the categorical imperative: "Never treat a human being as "mere means, but always as an end"[14]. It is not, however, mentioned just how greatly this formulation of the categorical imperative is apt to embarrass ethicists and indeed even ethicists who pronounced themselves deontologists. I offer some evidence for this. W. D. Ross remarks: the categorical imperative "has in fact great homiletic value; it is a means of edification rather than of enlightenment" ([24], p. 55). M. Singer says, it "has more of an emotional uplift than a definite meaning" ([25], p. 536); H. J. McCloskey writes: "It is not clear why

Kant thinks it to be wrong without exception to fail to treat persons as ends . . . nor is it clear why he thinks that it covers the whole of morality" ([16], p. 207). Very probably it is because of the "great homiletic value" that many ethicists like to refer to this second formulation, in order – deliberately or unintentionally – to exercise themselves in parenesis instead of in normative ethics.

Here I should not need to show that from Kant's second formulation of the categorical imperative alone the following moral standard cannot be deduced: "every innocent human being has at least the right not to be killed." But suppose that no human being should be viewed as property. Should that entail the proposition that no human being may be his own property, take himself as property? Is every kind of self-killing thereby morally wrong? Consider killing on request (e.g., the case of an incurably sick person); does he have a right to be allowed to die unhindered? The British doctor, Elliot Slater, comments on this and pleads for the moral permissibility of self-killing: "The life of a human being is his property and if we want to maintain that this is not the case, we would thereby be saying that he is a slave and not a free human being." He continues, "Slavery remains slavery even when loving and closely connected persons practice it" by refusing to accede to the request of the suffering for death. This train of thought is informative because Slater modifies a specious argument of the Christian tradition *against* self-killing into a specious argument *for* self-killing. The specious argument against self-killing: Only he can kill himself who is the master of his life; but the human being as a creature is not the master of his life, ergo. . . . Note that in the minor premise the word "master" is used in a different sense than in the major premise, so that the syllogism suffers from a *quaternio terminorum*. The specious argument *for* self-killing has the same major premise. Its minor premise reads: The human being is master of his life, for otherwise he would be a slave. Also, in this case, validity is compromised because "master" is used equivocally, as if the moral claim, whenever it assumes the form of a prohibition, made slaves of the human beings to whom it is directed.

III. PERSUASIVE DESIGNATIONS OR QUESTION-BEGGING WORDS

The way in which the word pair "freeman-slave" was used is faintly reminiscent of what for some time has been called persuasive designation or definition. As far as I know, this expression first appeared in Charles Stevenson's article, "On Persuasive Definition". But let us consider Susan Stebbing's use of persuasion in "Thinking to Some Purpose":

I shall adopt the arbitrary convention that "convince" is to mean "to satisfy by rational argument," i.e., by adducing evidence in support of the proposed conclusion. I shall confine the use of the word "persuasion" to mean "to bring about the acceptance of a conclusion by methods other than that of offering grounds for rational conviction" ([27], p. 76).

In a persuasive definition one refers to a person, a thing, or a fact with a linguistic expression which seems merely to describe, but which in reality already contains a normative or value judgment. A. Schopenhauer speaks of it in his *Eristic Dialectic* ("Artifice 12").

What a completely unintentional and impartial person would probably call "cult" or "public religious doctrine", someone else who wants to plead for it will call "piety" or "godlines", whereas an antagonist will call it "bigotry" or "superstition". Fundamentally, this is a subtle *petitio principii*: what one eventually wants to prove one puts into the word or definition in advance, from which it then apparently follows as just another analytic judgment. Among all artifices this is used most frequently, instinctively ([26], p. 14).

If one pays some attention to this, one will soon establish that Schopenhauer is all too right: "utilitarianism", "teleological ethics", and "ethics of responsibility" as defined by M. Weber [30] are sometimes used as synonyms in a purely descriptive sense. Yet when a Continental Catholic philosopher or theologian refers to teleological ethics as 'utilitarianism', it is highly probable that he disapproves of it, whereas it is virtually certain that by calling it "ethics of responsibility" he expresses his approval. In this case the terms "utilitarianism" and "ethics of responsibility", even though synonymous as to their descriptive meaning, are clearly antonyms as to their evaluative meaning. Word-pairs line "dynamic-static", "concrete-abstract", "existential-purely formal" admit of a purely descriptive use. But frequently they are used as value-terms in disguise. Who will be surprised when he is told that the true notion of natural law has to be a dynamic and not a static one, or that a purely formal concept of (Christian) faith is sorely defective, if not simply mistaken? Once more, such linguistic usage is interesting here only insofar as it constitutes a form of *petitio principii*. To cite Stebbing: "[there are] question-begging words, usually in the form of unpleasant epithets." As Mr. A. P. Herbert has said, "Give your political dog a bad name and it may do him more harm than many sound arguments" ([27], p. 163).

How do things stand in this respect with the just-cited characterization of utilitarianism from the pen of a non-utilitarian? Ultimately, for utilitarianism, all values can be traded off, including human beings. What evaluative meaning does the verb "to trade off" have? Is it the language of commercialization and business? The *Grosse Langenscheidt* has the entry:

> "to trade off = to sell [verhandeln], to barter away
> [verschachern]".

If the entry is to be correct, then we would have quite clearly a persuasive definition before us. The so-called careful consideration of values, i.e., the preferring of the higher or more urgent value, would be called a "bartering away" of the lower or less urgent value.

The verb "to trade off", "to barter away", suggests to me, by association, the notions of "embryo-transfer" and "surrogate-mother". It has been suggested that the *in vitro* fertilized ova can be transferred into the organism of the rent-mother instead of into the womb of the wife. Although there are a number of reasons for the transfer of an homologously fertilized ovum into the substitute mother, one could argue that these modes of collaborative reproduction cannot be ethically justified. In this case the process of procreation would thereby become "depersonalized" and "mechanized". This is most clearly exposed where somebody merely rents a substitute mother or a woman makes herself available simply for remuneration. It is not my intention to challenge such claims in his condemnation and final judgment. I ask only whether one who holds such a view does not already register his ethical rejection in the very characterization of those reproductive processes, and so employs 'question-begging" expressions. What about the expressions "loan-mother", "rent-mother", and "to rent a substitute mother"? In Duden's *Stilwoerterbuch* there is under "rent" [mieten]: "to rent a dwelling, a room, a stable, garden, a horse, an auto." Does not an expression like "rent-mother" already suggest that a person may be treated as a mere thing? Compare it with the German word "wet-nurse" [*Amme*]. According to Kluge's *Etymologisches Woerterbuch*, *Amme* "once stood for a mother insofar as the child is nourished by her." *Amme* does not have the connotation of commercialization. Why not? In English one frequently meets with expressions like "womb-hiring" and "womb-leasing". *Time* magazine (September 22, 1986) poses the following question: "Is the Womb a Rentable Space?" What kind of a question is that? Is it purely rhetorical or self-answering? Why do we not in the case of the wet-nurse also speak of "breast-hiring" or "breast-leasing"?

These expressions for their part find a correspondence in the criticism of the bourgeois view of marriage in the *Communist Manifesto*: "The bourgeois sees in his wife a mere instrument of production." Does not the figure of paranetic refutation appear in outline here which is always easily presented? It is partially explained from the double significance of the maxim: *do ut des* [give in order to get]. This maxim can serve, in addition, to characterize the behavior that corresponds to a *contractus bilateralis onerosus*. It then belongs to *justitia commutativa* and determines a form of ethically correct behavior. But *do ut des* can also be a designation of the prudential life adjustment of the clever egoist as it is described in *Luke* (6, 33): "And when you do good to them who do good for you, what sort of thanks do you have? Even the sinners do the same." This again makes clear how important it is to distinguish between the ethical predicates "good" and "bad" on the one hand, as well as between "right" and "wrong" on the other. One can do the right thing from bad motives. "Whether from dishonest motives or in sincerity, Christ is proclaimed, and that makes me happy" (*Phil*. 1, 18). Even if the motives which are supposed to move a woman to allow a child that is not her child to grow in her womb were to be rejected morally, it would not follow therefrom that her behavior is also morally wrong. Of course, conversely, it is also true that such behavior is not ethically correct, simply because her motive is nothing but unselfish love.

Let us turn to active euthanasia, typically called "mercy killing". J. Fletcher begins his essay, "Ethics and Euthanasia", with the following thesis: "It is harder morally to justify letting somebody die a slow and ugly death, dehumanized, than it is to justify helping him to escape from such misery" ([8], p. 100). Is this thesis not analytically or trivially true? It is only to be wondered which descriptive features let a death become "ugly", a human being become "dehumanized". Fletcher suggests that the ethical thinking articulated by this thesis is "humanistic" or "personalistic", "which puts humanness and personal integrity above biological life and function." Is not ethical thinking so characterized by definition the only correct and true ethical thinking? Passive euthanasia, i.e., "letting the patient go," is called "manifestly superficial, morally timid and evasive of the real point." In this way Fletcher lets us know that he himself and all who think as he does are "manifestly profound and morally courageous" ([8], p. 82). I wonder if those who morally approve only of passive euthanasia do not consider him and people like him "morally foolhardy". Once Fletcher, by the use of persuasive definitions, has justified active euthanasia with the appearance of direct evidence, a question inevitably presents itself to him: What reason can a human being, who is capable of thinking, have to deny what is immediately evident? Must he not appeal to *parenesis* in order to convince those who think differently: try not to be superficial, take heart, face the real point, and you cannot fail to admit that I am right!

I cite here the deontologist Hans Urs von Balthasar's criticism of contraception (except the rhythm method):

There remains a world of difference between the knowledge and utilization of the periods of bodily infertility by the married couple and the arbitrary, interfering fertility curtailment by contraceptive means. To many the difference appears small. And small it may be where the human being understands himself as the creature manipulating and discovering himself, as *homo technicus*. Where should limits generally be set for such a person in the loveless planning of himself? But for the person who thinks as a Christian the difference remains great.

The persuasive formulations are of course immediately conspicuous. Whoever uses artificial contraceptives acts arbitrarily and manipulates himself; he even plans without love. But von Balthasar does not leave it at that. On the contrary, he explains flatly that he who does not understand the ethical relevance of the difference between artificial contraceptives and the rhytmn method shows thereby that he does not think as a Christian, that he is of the opinion that the human being can and should invent himself. That sounds almost as if such a person denies the first article of faith, his own creatureliness.

Allow me to conclude my observations with some *paranetic* questions: Would controversies in the field of ethics not benefit if theologians and philosophers possessed a bit more humor? From time to time should they not recall that there is also a too disapproving and judgmental stance, which in the tradition is called *judicium temerarium*? Sometimes I wonder why Paul says in *Galatians* (6, 10): "We must do good to all and especially to our sisters and brothers in the faith." Should he not rather have written that we must be good and fair to all, *even* to our sisters and brothers in the faith?

Westfalische Wilhelms-Universität
Münster, Federal Republic of Germany

NOTE

[1] J. L. Houlden (in J. Macquarrie and J. Childress (eds.), *The Dictionary of Christian Ethics*) defines "parenesis" in the following way: "*Parenesis*: A word from Greek meaning advice, admonition, or exhortation (also spelled paraenesis). It identifies a form of ethical discourse or writing commonly employed in the literary analysis of biblical writings, e. g., *Matt.* 7; *Rom.* 12. The hortatory mode of ethical statement is characteristic of religious ethics. It presupposes a strong basis for duty and the need for improvement, and often a doctrinal framework in which the mere outlining of moral reasoning is inadequate."

BIBLIOGRAPHY

1. Balthasar, Hans Urs von: 1979, *Neue Klarstellungen*, 'Ein Wort zu "Humanae Vitae"', Johannes Verlag, Einsiedeln.
2. Barry, B.: 1973, *The Liberal Theory of Justice*, Oxford Press, New York.
3. Beauchamp, T. and Childress, J.: 1983, *Principles of Biomedical Ethics*, Oxford Press, New York.
4. Brandt, R. B.: 1959, *Ethical Theory*, Prentice-Hall, Englewood Cliffs, N.J.
5. Broderick, J.: 1961, *Robert Bellarmine: Saint and Scholar*, Newman Press, Westminster, MD.
6. Downie, R. S. and E. Telfer: 1980, *Caring and Curing: Philosophy of Medicine and Social Work*, Methuen, London.
7. Eibach, U.: 1976, *Medicine and the Dignity of Man*, Brockhaus, Wuppertal.
8. Fletcher, J.: 1973, 'Ethics and Euthanasia', in R. H. Williams (ed.), *To Live and to Die: When, Why and How?*, Springer-Verlag, New York.
9. Frankena, W.: 1963, *Ethics*, Prentice Hall, New York.
10. Hare, R. M.: 1963, *Freedom and Reason*, Oxford Press, London.
11. Hodgson, D. H.: 1967, *Consequences of Utilitarianism*, Clarendon Press, Oxford.
12. Jaeger, W.: 1973, *Paideia*, Oxford Press, New York.
13. Justin: 1971, *Dialogue with Jew Tryphon*, ed. J. C. M. Van Winden, Brill, Leiden.
14. Kant, I.: 1959, *Foundations of the Metaphysics of Morals*, trans. L. W. Beck, The Liberal Arts Press, New York.
15. Mabbott, J. D.: 1966, *An Introduction to Ethics*, Oxford Press, New York.
16. McCloskey, H. J.: 1969, *Meta-Ethics and Normative Ethics*, Martinus-Nijhoff, The Hague, The Netherlands.

17. Mill, J. S.: 1974, *Utilitarianism, On Liberty and Essays on Bentham*, ed. M. Warnock, New American Library, New York.
18. Moore, G. E.: 1966, *Ethics*, Oxford Press, New York.
19. Navreson, I.: 1967, *Morality and Utility*, Johns Hopkins University Press, Baltimore.
20. Norman, R.: 1971, *Reasons for Actions: A Critique of Utilitarian Rationality*, Barnes and Noble, New York.
21. Passmore, J.: 1961, *Philosophical Reasoning*, Gerald Duckworth and Co., London.
22. Plamenatz, J.: 1958, *The English Utilitarians*, Blackwell, Oxford.
23. Rawls, J.: 1971, *A Theory of Justice*, Harvard University Press, Cambridge, MA.
24. Ross, W. D.: 1939, *The Right and the Good*, Oxford Press, Oxford.
25. Singer, M.: 1961, *Generalization in Ethics*, Alfred Knopf, New York.
26. Schopenhauer, A.: 1974, "On Logic and Dialectic' in *Parerga and Paralipomena* trans. E. F. J. Payne, Oxford Press, London.
27. Stebbing, S.: 1938, 'Thinking to Some Purpose', Penguin Books, London.
28. Voltaire, F.: 1980, *Letters Concerning the English*, ed. and trans. L. Tancock. Penguin Books, New York.
29. Warnock, M., Chairperson, Committee of Inquiry into Human Fertilization and Embryology: 1984, *Report of the Committee of Inquiry*, London.
30. Weber, M.: 1965, *Politics as a Vocation*, trans. H. H. Gerth and C. W. Mills, Fortress Press, Philadelphia.
31. Wright, G. H. von: 1963, *The Varieties of Goodness*, Routledge and Kegan Paul, London.

JOHN LANGAN

MORAL DISAGREEMENTS IN CATHOLICISM:
A COMMENTARY ON WALLACE, SCHÜLLER, AND
THOMASMA.

Catholic approaches to ethical issues in general and to the problems of medical issues in particular have been both admired and criticized for their extensive use of philosophical categories and methods. The reasons for the Catholic reliance on philosophy in articulating ethical theory and in resolving ethical disputes come from different sources. First, from the long history of Catholic theologians from Augustine and Aquinas to Karl Rahner and Bernard Lonergan in our own time, who have been deeply engaged with philosophy. Second, from Catholicism's sense of itself as a universal church, as the carrier of a religious message that speaks to the hearts and the needs of all persons; this is a sense that was powerfully renewed at the Second Vatican Council and that underlies the recent pastoral letters of the U. S. Catholic Bishops. Third, from the Catholic understanding of the human person created as a free and rational being in the image of God, an image that is obscured, but not obliterated, by human sinfulness.

The dominant, but by no means exclusive, philosophical influence on the shaping of Catholic ethical thought for most of the last seven centuries has been Aristotelian scholasticism of which St. Thomas Aquinas was the most revered exponent. Father William Wallace, O. P., professor of philosophy at The Catholic University of America, carries on the Thomistic tradition of anchoring philosophical ethics in the philosophy of nature and more specifically in the philosophy of human nature. But he adds to this what he terms "a concordist view of Catholic theology and modern science." This gives his essay what his colleague, Dr. David Thomasma, of the Stritch School of Medicine at Loyola University, would call a Teilhardian breadth of vision. When he directs his attention more specifically at the ethical implications of his argument for delayed homization of the human embryo, Wallace acknowledges the inconclusive and tentative character of his argument. But it is significant that he looks to "further studies in reproductive biology" to resolve problems in the philosophy of human nature and not to the pronouncements of ecclesiastical authority, which he regards as more disciplinary than veridical.

79

Edmund D. Pellegrino et al. (eds.), Catholic Perspectives on Medical Morals, pp. 79–80.
© *1989 Kluwer Academic Publishers. Printed in the Netherlands.*

Dr. Thomasma raises questions about the bearing of Father Wallace's views on nature and creation on ethical dilemmas and implicitly brings us back to the disputes over the connections between "is" and "ought" which have marked so much moral philosophy in the English-speaking world in this century. Thomasma asks what help the scientific and theological perspectives that Wallace provides give us in answering questions about the termination of treatment and suggests that they turn our attention from the domain of interpersonal interactions that we should be considering.

The philosophical sources that engage the attention of Professor Bruno Schüller, S.J., of the University of Münster, are significantly different. Like many of the analytic moral philosophers of the English-speaking world, Schüller is concerned with the tasks of separating moral argument from parenetic discourse or moral exhortation and of minimizing reliance on persuasive definitions to do the work of serious argument. Schüller's work here and elsewhere constitutes an important point of connection between analytic moral philosophy and Catholic moral theology. Schüller also draws on biblical sources in a disciplined and scholarly way which recognizes the importance of literary genus in determining just what the biblical contribution to our understanding of moral issues is. John Langan, S.J., of the Woodstock Theological Center at Georgetown, uses an approach akin to Schüller's to explore the connections between agreement in resolving moral cases and disagreement in moral theory. Both are opposed to condemnations of teleological reasoning on moral issues and to proposals to interpret Catholic moral theology on medical and bioethical matters as a series of deontological prohibitions.

Philosophy serves as a vital link between the questions put by our developing technology and our often troubled and anxious culture and the norms and values commended by the Church both in its practice and in its hierarchical teaching. Yet it is itself a diversified and changing area of inquiry. No presentation of Catholic medical morals would be adequate if it neglected to consider both the contribution and the controversy that philosophical reflection brings to the issues of bioethics.

Georgetown University
Washington, D.C.
U.S.A.

PART III

THE THEOLOGICAL FOUNDATIONS

JOSEPH FUCHS

"CATHOLIC" MEDICAL MORAL THEOLOGY?

At the end of his 1975 Père Marquette lecture on the "The Contributions of Theology to Medical Ethics" [2], J. M. Gustafson comes to the
conclusion that these contributions are not particularly great. The
volume *Theology and Bioethics: Exploring the Foundations and Frontiers*[3], edited by E. E. Shelp, confirms this conclusion. It follows that
we can hardly expect greater things when we deal here specifically with
the contribution of *Catholic* theology to medical ethics. Nevertheless,
reference is made frequently to Catholic medical ethics, presumably
because Catholic moral theology and the Catholic Church have dealt
more frequently in the past with questions of medical ethics than have
other religious institutions.

I. MEDICAL ETHICS

Catholic medical ethics has generally presented itself as a philosophical
ethics: its reflections, its principles, and its reasonings differ hardly at
all, in a formal sense, from those of a philosopher. But if it is to be
specifically Catholic, it cannot simply be philosophical, but must proceed from the Christian and Catholic faith. It is obviously possible for
Catholic faith to give support to such an ethics without displaying it; it is
indeed perhaps possible that the moral theologian in his endeavors is
not explicitly aware of his faith as the basis of his endeavors.

This presupposes that, although the Catholic faith and Catholic
theology wish to be significant for a Catholic medical ethics, this latter is
in principle possible – at least as normative ethics – without explicit
reflection on the faith. This in turn means that it is taken in principle to
be possible to arrive at this ethics on the basis of a common dialogue
between the believing Catholic and the non-believer.

Faith itself, which must serve as the basis and starting-point of a
Catholic medical ethics, requires a moral justification that is (logically)
antecedent to itself and hence, as such, cannot derive from faith alone.
It follows that medical ethics cannot appeal to a faith as its ultimate
basis. In medical morality the self-understanding of the human person

83

Edmund D. Pellegrino et al. (eds.), Catholic Perspectives on Medical Morals, pp. 83–92.
© *1989 Kluwer Academic Publishers. Printed in the Netherlands.*

as a moral being (logically!) precedes every faith and every believing reflection. This means that every medical-ethical reflection that derives from faith occurs within a human self-understanding that (logically!) precedes faith; the immediate application of this is merely that the human person understands himself as a moral being, and therefore necessarily understands medical conduct and action also to be moral conduct and action. In other words, because medical problems are human problems, they are always moral problems, too; their moral quality does not derive in the first place from a faith.

This is true also of the content of medical problems. It is not the case that they have a moral solution because a God in whom we believe has ordained things in a particular fashion. The well-known formulation that everything is allowed if there is no God as starting-point, is false. Nor is it true that medical conduct or action is right only because a God has willed it thus. Rather, the question of what medical conduct and action correspond to humanity must be solved on the basis of the self-understanding of the human person.

Accordingly, a medical ethics deriving from Catholic faith or Catholic theology can have no value if it is unable to accord with human self-understanding. Without such correspondenced, an ethics based on faith or on theology could not be understood humanly, and could therefore not be understood as morally obligatory. It would be humanly meaningless.

Medical ethics is thus an autonomous ethics – in the sense that it is not merely imposed on the human person from outside himself. Catholic Christianity *per se* is therefore not an original source of replies to the questions concerning right medical conduct and action. If one is not afraid of the concept of moral natural law (understood not statically, but dynamically), this is the right category for medical ethics. Understood theologically, this means the created being of the human person in its orientation toward the future; both something given, to which one must correspond and a future which must be discovered. It is the human person's task to seek and to find. It is his tragedy that he can go astray in this and arrive at false solutions, just as it is humanity's tragedy that different people can arrive at different solutions to the problems that present themselves in this search.

Plato and Hippocrates did not understand humanity with its value and dignity in the same way. Unlike Hippocrates, Plato saw humanity in its usefulness for the state as a whole, and believed he knew which human life should be procreated, born, and kept alive. Today, too, the question of "Who may live?" is posed, though rather from an individualistic or indeed egotistical point of view. A similar question arises from the point of view of the quality of life, in the case of very young or very old human life, or utterly weak human life; what should one risk and undertake for such life? Besides this, when is human life in fact the life of a human person, and what is the consequence of the doubt that is indicated here? And finally, who is to decide how a human life is to be disposed of?

This does not mean that the God of faith and thus faith and theology itself are irrelevant to human attempts at solutions. The God of creation and redemption wills the search to be a search for *true* solutions, i.e., solutions that truly correspond to humanity. An utterance of God – e.g., through his prophets such as Moses or through the Bible – insofar as it genuinely intends to address medical ethics, can only be an utterance that is in accord with the human person: it follows that it is fundamentally comprehensible to the human person and that the human person himself is capable of thinking it. This does not mean that such a word of God is superfluous; it has a maeutic function for the human person in his reflection and his search, and gives him additional certainty *vis-à-vis* what he himself discovers; it can indicate an orientation or a direction in which a good solution to the problem must be sought and found.

II. THEOLOGICAL MEDICAL ETHICS

Medical ethics is theological, and hence, Catholic-theological ethics, if it proceeds from faith, i.e., from the Catholic faith. This faith is ultimately not the assertion of the truth of certain faith-propositions, but an act, in the depth of the person, of giving and entrusting oneself to the God who reveals and imparts himself to us. Naturally, no concrete ethics – and therefore no medical ethics – can be developed out of faith understood in this way.

God's self-revelation and self-gift to man seek man's salvation, i.e., his life in God instead of his death. The self-giving of the Son is a self-giving "for us" – for the others. Jesus' self-giving in his earthly life is a pouring out of himself "for us" – for the others. His life is a life for the "salvation" of the world, i.e., a life and activity and self-giving for our life with and in God – instead of death. His concern for the conduct of our lives is that we, like him, should be "just" instead of sinners who are egotistically closed in ourselves, and hence that we should be open to all that that is right and good, to the others and to God. Jesus' way in the world of human persons was one of doing good, helping, and healing – all an image of the God who reveals and imparts himself for our salvation. Here medical research and the endeavors of doctors for the good of men are called upon: where or how do moral norms exist for them?

Our act of giving and entrusting ourselves to the God who reveals and imparts himself to us is, at the same time, an acceptance of our own self. For its part, this self-acceptance which takes place in the depth of the person is the acceptance of the task of understanding this self and developing it in the direction of self-realization. Thus, the human and philosophical attempt to understand and develop oneself *vis-à-vis* medical problems is ultimately an attempt that derives from *faith* and is propelled by faith. This is so even when the believer does not reflect explicitly on the deepest basis of his search. What seems to be human philosopohy is, therefore, the attempt in faith to think out and translate this faith into the concrete medical realm of human reality. It is in this sense that the medical ethics of the Catholic who holds on to his faith in his moral reflection is a Catholic ethics, even though it works with the intellectual tools of the philosopher. This, however, is only a partial truth, because living faith also signifies an interior disposition to seek the truth, the whole truth, and nothing but the truth, and such a disposition, present in the basis of faith, has primary significance for the seeking and discovering of medical moral norms.

Medical ethics of this kind serves also to make explicit Catholic theology. For the faith is also made categorial in faith-propositions, in which it is affirmed how God is in Jesus Christ *vis-à-vis* man, man who has become a sinner, and every man. At the same time, much is also affirmed about man in such propositions. Here, for example, is made clear what the ultimate dignity of the human person, and every human person, consists in; it becomes manifest that this dignity belongs to every human person, and that ultimately therefore all human persons have the same dignity. Should such affirmations create difficulties in the human-philosophical endeavor, the believer can derive their truth from his faith and a corresponding theology. Should such anthropological affirmations not exist, or fail to convince, in the search for norms of a medical ethics, then the faith can become the "light of the Gospel" – to use the words of the Second Vatican Council. In this way, faith can also be significant both in terms of the contents of medical ethics, and for the doctor. It remains, however, true that the faith and a theological anthropology only offer help and support for the establishing of a medical ethics, and that they cannot replace human and philosophical reflection, seeking, and discovery. But the danger of mistakes remains, as does the possibility that the many who seek the elements of a Catholic medical ethics on the basis of their faith may not all arrive at the same result. This is unavoidable, since the various elements of a medical ethics are not revealed, and hence are not the object of faith: they must be discovered "in the light of the Gospel" in the human search – but the search remains human.

One should note that what has been said holds true for a normative ethics (moral rightness), but not only for this: it is true also of the morality of virtues (moral goodness). If one understands and accepts the human person – including the patient – in the ultimate reason for his dignity, one will arrive at an attitude and a conduct that would perhaps otherwise have been absent. One who understands human life – and this means, in fact, human existence – in this way will more easily discover conduct and a correspondingly right action, than one who must discover his path without a correspondingly solidly-based ethics. Here we may think of the example posed by the complicated problematic of euthanasia.

III. CATHOLIC TRADITION

It is well known that the Catholic Church and its theology attach great weight to the Christian tradition and orientate themselves accordingly. This is understandable, given that revelation and faith are given in advance: it is on the basis of revelation and faith that theology is possible as a reflection on the faith. Tradition, however, has a different meaning for theology than for faith; for theology, as human reflection, is conditioned by human elements of the reflection and can therefore in certain circumstances arrive at diverse theologies. The significance of tradition for ethics, including medical ethics, is even more problematic.

There are fundamental revealed and believed truths that are relevant to ethics. On this basis, it is established that God has taken up all human persons into his love, that all have become sharers in this love, that the dignity of human persons lies most deeply therein, that all human persons have become sharers in this dignity in the same manner, that all must acknowledge this dignity in all persons, that therefore no one may close himself egotistically in himself, that he must be fundamentally open to all persons: such truths are established as the foundation of every Catholic ethics.

It is a different matter when such truths are to be made concrete. Here revelation and faith no longer say the last concrete word, for here human reflection begins. Here the reference to the fundamental dignity of the human person no longer suffices; the more concrete question is pressing: how a particular act, observing human values, does or does not preserve the dignity of the human person – and this question is not answered by revelation and faith. Nevertheless, great value is attached to tradition in Catholic moral theology. Once it was thought that the Christian past always developed its concrete ethics "in the light of the gospel" and believed that it certainly found itself, in the ethics developed at each period, on a path in harmony with the gospel and thus with the faith: in addition, it was thought that the Holy Spirit is always with the Church of Christ in her seeking and discovery, and that he guarantees for her teaching at least the presumption of truth.

Today, people of various outlooks on the world – including Catholics – make a somewhat different judgement with regard to such moral solutions of a Christian tradition. Indeed, Catholics have given different answers to individual questions at different periods in the history of Christianity. We ask: where then is the tradition that is to help us? Do we today perhaps know human reality in a partly different way from that of earlier periods? Do we perhaps make human judgements differently? Do we perhaps make a different interpretation of human situations from a moral point of view? If so, then a different human evaluation, and a different moral norm-setting must occur. This holds true also for questions of medical ethics.

We are confronted here with a decisive problem. If neither faith nor dogmatic theology can offer us concrete solutions to concrete moral problems, and can do no more than preserve the light of the most general principles, then the concretization of such principles in the concrete circumstances of life must take place in another way, i.e. by means of the human evaluation and moral interpretation of the given human realities by the Christian as a human person – hence in a way that is fundamentally available even to the non-Christian human person, even if the Christian acts always "in the light of the gospel". This is the way that we are accustomed to call the "moral natural law".

Given the lack of corresponding biblical affirmations, the Christians of the first centuries often adhered to evaluations and norms given by pagan philosophers – e.g., Stoics and Gnostics – in their search for moral norms, provided that they believed that they were able to find no incompatibility between the concrete moral norms given by the pagan philosophers and the fundamental attitudes of Christians. For example, the Christians generally took over the doctrine of *apatheia*, which had as its aim a conduct that was not at all guided by passions; this was held to be a very high moral ideal. One of the consequences of this was that in Catholic moral theology in general, for many centuries – even if not today – sexual intercourse within marriage that was not concretely motivated by the will to procreate new life was held to be (at least venially) sinful – clearly, a doctrine that could acquire importance within medical praxis.

But this is only one single example of a one-sided and rather mistaken moral interpretation of a human situation. More significant is the way in which the Stoics taught that the given nature of the human person should be understood as the norm of moral conduct. The Stoics held the human givenness, called nature, to be divine in a pantheistic sense. It follows that what nature indicates is understood as the divine will: if nature shows the growth of a beard as God's will for man, then man must not shave. If nature gives us to understand that the only goal of sexual activity is the procreation of new life – as we mistakenly thought – then all sexual conduct that does not correspond to this objective is to be judged immoral. Christianity, which understood human nature not pantheistically but as the work of the Creator, took over this teaching, thus "Christianised" from the Stoics. This was true not only of the teaching about marriage, but in other spheres of life, too. In the middle of this century, the permissibility of the transplantation of an organ from a living organism was still a much discussed question; for it was argued that nature (the Creator) had, for example, willed this particular kidney for this particular organism and not for another. One can readily suppose that such an understanding also plays a role in problems of the most modern bioethics. The traditional teaching about the natural law is a treasure of Christian moral theology, but it is not to be taken over blindly. As a norm of moral conduct, it must be brought into relationship with "right reason", which has to achieve the balance between various human goods and values.

We may draw attention to other problems of Christian tradition. One of the jewels of the entire Christian tradition is its insistence on the value and dignity of human life. But, even on this fundamental point, tradition had a different understanding during the first centuries than in later periods, e.g., regarding the question of killing in war, which was excluded in the earlier period. In order to preserve the unity of Christians in the one Christian faith, tradition did not always guarantee the protection of life – that is, unlike today, it allowed killing. At all times Christian tradition has been the protector of newly born life; i.e., in rejecting both abortion and infanticide. When both problems are discussed afresh today in wholly other circumstances and possibilities, can one simply appeal, without further question, to a tradition that, in different circumstances, came to a judgment about the protection of human life? This question must at least be posed.

In short, the tradition of Christian moral theology developed generally in "the light of the gospel", but as a Christian-human attempt at a concretization of the principles of the gospel. This tradition is not valuable for us because Christians could consider it for long periods as corresponding to their fundamental Christian outlook. But since it was always also a human attempt, and as such could stray from the truth. Catholic moral theology must not take it over blindly in all points. Only in this way can it serve today as a Catholic tradition.

IV. THE CHURCH'S MAGISTERIUM

It is well known that the Catholic Church attaches great importance to its tradition, even in questions of ethics; it is even better known that it officially prescribes a Catholic moral doctrine, and so likewise a Catholic medical ethics, as its obligatory doctrine. This is connected to its understanding of the Church's tradition: within the Church's fellowship with its tradition, there is an official and thus valid exposition and interpretation of this tradition. Further, inasmuch as tradition is always understood in an ultimate relationship to the doctrine of the faith of the Church, the official teaching in questions of ethics, which is upheld in each period and today as valid – e.g., in questions of medical ethics – is seen as standing in an ultimate relationship to the doctrine of the faith of the Church. It is precisely for this reason that the Church, as the guardian of the faith, intervenes in an official way in questions of moral theology, including those of medical ethics.

This is of interest inasmuch as the Church's doctrine of the faith does not as such contain the many concrete moral questions, and therefore does not generate them out of itself: they do not belong to the treasure of the Church's faith. There is no special knowledge about the essence of sexuality, and correspondingly about right sexual conduct, in the treasure of the Church's faith, nor is there special knowledge about the beginning of human life, about the reality of embryos, or about the nature of genes. Nevertheless, corresponding knowledge about these matters is presupposed in moral affirmations about human behavior with regard to these realities; for general moral principles do not suffice. The Church, however, insists on its magisterium in questions of medical ethics. The many interventions of Pope Pius XII are still well remembered, and the interventions of the present Pope are well known.

In this way, the magisterium of the Church shows itself as the competent authority not only in questions that count as revealed, but also in non-revealed questions of natural moral doctrine. This claim is explicitly asserted, as was done several times during the Second Vatican Council. There are, however, formulations both in the Second and in the First Vatican Councils to indicate that questions of natural law are not bound to to the magisterium in the same way as questions that obviously have to do with Christian revelation. Here it is clear that not all questions have been resolved yet ([1], pp. 57–67).

Nonetheless, if one is questioned about doctrine in matters of medical ethics, one refers to papal and episcopal statements. Many Catholics feel bound for their instruction and their praxis by such statements; this is also true of biologists and theologians. But it does not apply to all moral questions. It belongs to Catholic tradition and also to the teaching of both Vatican Councils that only a few ecclesiastical pronouncements have to count as "infallible". The "non-infallible" statements include the teachings which are not found in the treasury of the Church's faith, but for example, like concrete moral norms, derive only from human reflection (from moral natural law). The Church's magisterium does indeed emphasize very explicitly that one must adhere religiously to such teachings, too; if questioned, it would insist on this. But dissent would not be absolutely excluded; this is clear from the totality of the proceedings at the First and Second Vatican Councils.

The faithful of the non-Catholic Christian churches of the West do not feel bound in the same way as Catholics by an ecclesiastical magisterium or by a Christian tradition in questions of ethics, including those of medical ethics. In part, this is because these churches do not understand themselves to be hierarchically structured in the same way; consequently, the church's leadership is not so important and so conspicuous in them. They do, however, make statements, both for their own believers and for all men of this world. These leaders believe that they have a mission to do this, believing that they should give light from the resources of the Christian faith; but they issue fewer requirements and commands in dealing with their faithful. Like the church leadership, the pastors and the theologians of these churches do the same, though on a different level.

This is so, although the theology of the other churches makes less use in its work of the concept of natural law than does the Catholic church. They believe, rather, that one can and should speak more directly of God's will. This is often not a revealed will of God e.g., in the Bible, but rather insight gained in the Holy Spirit, which the believer possesses – together with a willingness to take into account, in great freedom, the directives of the pastors, the theologians, and the church leadership. To take an example of some theologians: many hold the commandment "Thou shalt not kill" to be an absolute divine requirement, but see that in particular cases the requirement is too much for the weak and sinful human person. They deduce from this in the Holy Spirit that God, too, does not insist: the act remains a sin, but the sinner is certain of pardon. In other words, God justifies not the act, but the sinner. Catholic theology thinks otherwise: when one reflects precisely in terms of natural law, in the Holy Spirit, it is seen that God's requirement is not so extensive as it could at first sight seem to be. The requirement of God, thus acknowledged, is however, absolute; there is no need either of a justification of the deed or of a justification of the sinner.

Pontifical Gregorian University
Rome, Italy

BIBLIOGRAPHY

1. Fuchs, J.: 1984, *Christian Ethics in a Secular Arena*, Georgetown University Press, Washington, D.C.
2. Gustafson, J. M.: 1975, *The Contributions of Theology to Medical Ethics*, Marquette University Press, Milwaukee, WI.
3. Shelp, E. E. (ed.): 1985, *Theology and Bioethics: Exploring the Foundations and Frontiers*, Kluwer Academic Publishers, Dordrecht, Holland

LISA SOWLE CAHILL

"THEOLOGICAL" MEDICAL MORALITY? A RESPONSE TO JOSEPH FUCHS

Joseph Fuchs is a skillful and experienced articulator of the contemporary meaning of "natural law" in the Roman Catholic tradition of ethics. Over twenty years ago, he contributed a major discussion of the question whether "natural law" is a purely philosophical concept in Catholic tradition, or whether it can be grounded theologically [4]. The fact that the conversation has continued for decades indicates that it is of key importance to the intelligibility of natural law ethics as Christian ethics, and also that the question is one not easily settled. I will devote my comments to four issues suggested by Fr. Fuchs's present treatment. An initial concern already has received our attention: (1) the premises and method of the particular philosophy in mind when we address the relation of philosophical thinking to *Catholic* medical ethics. The next three issues are pertinent specifically to the theological foundations – and ramifications – of natural law philosophy as interpreted within Catholic tradition. These are: (2) the relevance of Scripture to Catholic ethics; (3) the authority of the Church as interpreter of a philosophically-based ethical tradition; and (4) a proposal regarding the interdependence of theological and philosophical resources in Catholic moral reflection.

I. "NATURAL LAW" PHILOSOPHY IN MODERN CATHOLIC TRADITION

As other contributors indicate, the philosophy which forms so integral a part of Catholic moral theology is a late twentieth-century version of Aristotelian-Thomistic "natural law." Insofar as this philosophy is premised on universal human inclinations, values and virtues, its substantive conclusions are said not to have any specifically theological foundation. In fact, it is a key characteristic of the Catholic natural law tradition to deny that any religious commitment or any theological anthropology is a prerequisite of genuine moral insight. As Fr. Fuchs observes, natural law morality is congruent with "human self-understanding [3]. An asset of this approach in our century is that it has served to establish a community of moral discourse accessible in principle

93

Edmund D. Pellegrino et al. (eds.), Catholic Perspectives on Medical Morals, pp. 93–102.
© *1989 Kluwer Academic Publishers. Printed in the Netherlands.*

to members of different moral and religious traditions. It represents a
commitment to some essentially invariable human values, to the perspi-
cuity of these values to reasonable reflection, and to evaluation of
particular moral decisions and acts with due seriousness and precision
("casuistry"). Moral discourse is said to proceed from faith primarily in
the sense that the universality and knowability of the moral law are
grounded by a divine Creator who made humanity in his image and
endowed humans with intellect and freedom.[1] Neither Scripture, nor
"faith" understood more generally, is a sufficient basis for the direct
derivation of specific moral precepts. However, it must be held in mind
that even while explicating the "law of nature", Christian authors
operate within the general context of Christian commitment and theology, and so any simple separation of faith foundations from substantive
natural law conclusions is bound to be artificial. To the extent that
Christian natural law representatives have been insufficiently conscious
of their own historical roots, specific moral proposals derived within a
specific – and thus necessarily circumscribed – cultural, philosophical,
and religious tradition have been put forward authoritatively in Catholic
teaching as the rational and immutable requirements of human nature
as such. A task for Catholic natural law thinking today is to examine its
philosophy more directly in relation to the religious and theological
aspects of the tradition, and to consider the degree to which those
aspects have helped to shape the "natural law" conclusions.

II. SCRIPTURE AND NATURAL LAW ETHICS

In a document of the Second Vatican Council, the *Decree on Priestly
Formation* (1965), it is urged that "Special attention needs to be given to
the development of moral theology. Its scientific exposition should be
more thoroughly nourished by Scriptural teaching [9] ." Writing even
before the Council, Josef Fuchs was among the first Catholics to mount
the natural law method on an explicitly biblical foundation, appealing
especially to *Romans* 1:18–32, and also to the biblical doctrine of
creation, including the Prologue to John's Gospel ([4], ch. 2). Fr. Fuchs
stands in a respectable tradition of Catholic authors from St. Thomas to
the present, who have resolved the problem of a specifically Christian
morality in favor of a general concurrence of NT and natural law, with
biblical religion or faith primarily in the role of motivator, and natural
reason in the role of practical guide.[2] However, when one places the

biblical materials in the foreground and attempts to develop an ethics first and foremost on that basis, questions arise as to the sufficiency of this solution. An obvious question is whether the natural law tradition ever has dealt in a fully convincing way with the radical nature of Christian discipleship, and its transformation of the believer and the believing community. The Catholic biblical scholar, Raymond F. Collins, recently has contrasted the usual approaches to the question of the relation of Scripture to ethics with the one which seems to be suggested by study of the Bible in its own right. Usually, the focus either is on those biblical materials which appear to have specific ethical relevance – in which case a negative answer results due to the problems of method in applying such texts today and to the difficulty in identifying any specific precepts which clearly are unique to Christianity; or on a generalized understanding of Christology or the teaching of Jesus – in which case a positive view of the relation results, but one which is in many ways abstracted from the concrete perspectives of the NT authors. Emerging as a biblical exegete more out of sustained contact with biblical points of view than out of a concern with ethical theory and a universal morality, Collins "would rather reverse the question and consider how the biblical authors treat ethical issues ([1], pp. 23–24)." Recent historical studies of the biblical materials have made it more clear than ever before that the communities that produced the texts expressed the meaning of Christianity from within views of the world which differ in significant respects from our own. A chief difference, for instance, is the fact that the earliest communities expected Jesus to return as Lord within their own lifetimes. Naturally, this eschatological expectation colored their ideas about the behavior expected in the present in areas both of personal morality and of social responsibility. Today, it is to be asked whether NT eschatology has any remaining significance for Christian ethics, and, if so, what it is. Thus introducing natural law philosophy more completely to biblical resources cannot be a simple matter of matching the conclusions of philosophy against those of religion, in order to determine areas of congruence or incompatibility. It also includes the task of trying to appreciate the meaning of a Christian morality from within the biblical worldview itself, even if that perspective and its consequences for morality are very different from the ethics to which we as Catholics are accustomed. While natural moral theologians endeavor earnestly to demonstrate the basic compatibility of Thomas' ethics and the gospels, biblical interpreters are posing the question whether the meaning of

Christianity has been understood at all if ethics is taken as an autonomous object of interest, rather than as part and parcel of discipleship ([6], p. 125).

Many exegetes today would agree that a basic NT insight is that righteous living proceeds from conversion and "singlehearted devotion" to the Kingdom of God (1 *Cor* 7:35). We find in the NT little striving for an objective, humanistic moral perspective; emphasis is on participation, involvement, spontaneity, and Spirit-inspiration, transfused by urgent eschatological expectation. Collins remarks that "personalism" is linked to the ethics of the Sermon on the Mount (*Mt* 5) in that both see the moral life as emerging from and consistent with the "being" of human persons. However, from the standpoint of the Sermon, ethics proceeds from one's being as a Christian, as a "member of the Kingdom", ([1], p. 227). The moral question of the NT is not first of all, "Can non-Christians arrive at the same ideals?", nor "Is the content of Christian ethics "different" from that of the pagans?" The point of interest is not the convergence of particular moralities in a common human ethics. It is, rather, the nature of the Christian life in its own right as a life of discipleship, though that life certainly does not preclude agreement with the moral conclusions of others. St. Paul in particular makes ample use of moral wisdom, some of which we, with hindsight, may see as limited. Further, we find in the NT virtually no interest in a detailed, practical casuistry; moral exhortations are illustrative (rather than specifically definitive) of what discipleship demands. This leads to the question just how these exhortations are to function for us today. The biblical illustrations of discipleship do not bring us directly into the realm of medical ethics, but they certainly seem to overturn preconceptions about the ordinary meanings of natural justice and rights and duties (*Mt* 5:38–48). As depictions of the Kingdom, such illustrations are not so much "exhortations" as "indicatives" which have at the same time a certain "imperative" and also judgmental force. If Christians are not even now making present the Kingdom of God in the way illustrated, they should be, and will be held accountable.[3] Perhaps the distinctive characteristic of a biblically-grounded ethics is not its content, but the fact that this content is presented in terms of urgent, obligatory demands. Even love of enemies and self-sacrifice are not presented as distant, supererogatory ideals, but as constitutive of discipleship (*Mt* 5:48–38).[4]

Furthermore, if Catholic moral theologians are to engage in biblical interpretation as an endeavor central to their enterprise, then they must be prepared to deal with many additional hermeneutical problems posed by recent biblical exegesis and interpretation. These include historical-critical questions such as the definition of the "canon" and of its authority as such; of the relative authority of those parts of the Bible which are not readily harmonized; of the criteria of that authority, particularly if it varies from part to part; and whether factors such as historically earlier or later authorship, the influence of socio-cultural factors on biblical authors, or their use of philosophical images or categories, can diminish the authority of a text.

III. NATURAL LAW PHILOSOPHY AND CATHOLIC TEACHING AUTHORITY

The authority of a religious body to interpret the principles and conclusions of the natural law is not self-evident on a philosophical basis. This authority depends on claims which that body makes by virtue of its special religious foundations, prerogatives and responsibilities. An obvious tension – not to say disjuncture – results in the relation of the role of natural law ethics as philosophical ethics to its role as a system of warrants for what are finally the claims of a religious tradition. While the magisterium of the Church understands itself as having authority under the guidance of the Holy Spirit to interpret the moral implications of philosophy, scripture and contemporary experience, it does not see itself as equally subject to the critique of any of these other sources. It is difficult to avoid the further question of exactly how the Catholic Church, as a *religious* and *theological* community, is to function authoritatively in the philosophical definition of moral obligation. The question becomes all the more complex to the degree that the magisterium itself operates, not out of "philosophy" in the abstract, but out of twentieth-century Western Thomism. The gradual development of that philosophical school has occurred in tandem with the Church's concern to address the religious, theological, ecclesiastical, and social problems that have impinged on its history over the past seven centuries. When authoritative moral proposals are made, it must be asked whether they can stand on philosophical warrants, and if not, whether the theological reasons backing them have been explicated adequately.

Since the Second Vatican Council in the 1960's, a new perspective has been gained on natural law as the approach of a historical religious community under the auspices of a formal *magisterium*. Increasingly, natural law moral precepts have been supported with religious, especially biblical, symbols and language. Perhaps more significant in terms of potential impact on the specific conclusions of Catholic morality are the insights that knowledge of the natural law is mediated through *experience*, and that the evaluation of any sort of individual act must take account of the *social* context of that act. Pope John XXIII suggested a renewed reading of natural law through the lens of present, concrete experience when he called Catholic theology to an encounter with the "signs of the times."[5] More recently, John Paul II has made persistent appeals to concrete experience in his audience talks on sexual ethics, especially in the context of married love.[6] In the modern papal social encyclicals, there has been a strong tradition of applying natural law in the context of common good, mutual rights and duties, and social justice. Recently, valuable links have been drawn between this social justice tradition and those areas of medical ethics which previously had been construed as "personal" ethics. A good example is Cardinal Joseph Bernardin's metaphor of the "seamless garment" which unites war, capital punishment, euthanasia, abortion, health care for the elderly or indigent, and a decent standard of education and employment.[7]

These new directions raise important questions that will be crucial for the shape Catholic medical ethics is to take in the future. The turn to *concrete experience* raises the question of the criteria of *valid* or *normative* experience. The emphasis on actual, particular experience seems to suggest that special attention should be given to persons who participate in the experience under evaluation, and that the witness of those persons should have some impact on the resulting normative conclusions. One is led to ask whether or in what way the pontiff's reflections on the experience of the married does or should reflect dialogue with married persons. In medical ethics in particular, concrete experience bears the mark of the "contingency" that Thomas Aquinas saw as the condition of the operation of the practical reason.[8] If, to the extent that variability characterizes a certain realm of moral obligation, the articulation of the requirements of the natural law becomes less certain or more variable in "matters of detail" [11], then enhanced recognition of the experiential base of natural law ethics may have important consequences for the adaptation of the conclusions of medical ethics about, for example, care for the chronically or terminally ill, reproductive technologies, human experimentation, and the fair distribution of basic health care. As Fuchs has said, natural law ethics today is to be understood "not statically but dynamically" ([3], p. 3). In these same areas, we see the important relation of *social context* to any evaluation of individual acts. Should decisions about care for the very ill or about the utilization of techniques to remedy infertility reflect the problem of scarcity in resource allocation? If so, in what way? How do such bioethical evaluations reflect or feed into larger social attitudes toward the importance of individual lives, the role of medicine on our culture, the nature of parenthood and the family, the roles of women, international economic justice, and so on? What are the institutional responsibilities of Catholic health care facilities? These questions appear particulary difficult when, realizing that the natural law tradition with which we are most familiar has developed in the prosperous, post-Enlightenment West, we attempt to take into account the experiences of persons and communities cross-culturally, and to consider what effect these considerations will or should have on the way we institutionalize moral authority in our theological-ethical tradition.

IV. CONCLUDING PROPOSAL

The Catholic natural law tradition has always in fact relied on several interdependent sources of moral insight, including concrete human experience, philosophical reflection, religious images, theological interpretation, and magisterial proposals which attempt to balance out the tensions in all of the above. These sources of moral thinking must continue to inform one another. It is when one element in the process is translated as an absolute point of reference that distortion occurs. A fundamentalist biblicism results in simplistic "proof-texting"; an abstract elevation of "natural reason" results in the rigid, ahistorical, and individualist ethics of which we can find some examples in the old moral theology "manuals"; the absolutization of tradition for its own sake results not only in authoritarianism but in defensiveness; total immersion in the exigencies of immediate experience produces the subjectivity and relativism to which legalism is the natural backlash.

Among the most valuable theological and ethical resources which Catholic tradition has to offer to biomedical ethics is its recent social teaching. Early on, contributions to Catholic social tradition recognized in a clear way that solutions to complex problems emerge out of communal discernment and yield norms of action that will demand refinement in the process of application to specific and to an extent unforeseen social realities. The late nineteenth and twentieth century encyclicals on social justice (beginning with *Rerum Novarum*[9]) have been able to sponsor successive creative responses which shift substantially in their practical recommendations, even while remaining within the same essential parameters of human social cooperation. More recent teachings, such as the U.S. Bishops' pastoral letters, represent a rich cooperation of biblical themes ("Kingdom of God," "preferential option for the poor"), philosophical categories and values (justice, common good, rights, duties), experience-based analysis of the concrete realities with which morality is concerned (health care, employment, the arms race), and respect for precedent tradition (reference back to previous magisterial analyses of social conditions). The ambiguities arising from such a multiplicity of reference points have not been resolved fully. Perhaps it will be impossible to do so, so long as the Catholic natural law tradition aims, as it sometimes is put, to be both "authentically Christian" and "authentically humane." Yet, the explicitly dialogical character of the models recently provided is an advance on a more narrowly construed natural law method not genuinely representative of the complex and multi-faceted process by which Catholic Christian moral evaluation and teaching evolve.

The new, dialogical model of teaching may represent a revised conception of the "natural law" itself, one more attuned to the historical and cultural particularity of any philosophical viewpoint. Perhaps it is a misconception to view the natural law philosophy of Catholicism as providing access to a realm of human knowing and discourse in which the members of particular traditions can be provisionally cut loose from their roots in special religious and cultural experiences. Perhaps there is no such privileged realm "beyond" the particularity of persons and communities. As Gerard Hughes has observed, relativism may here have a point. At the same time, however, relativism is only partially successful. The characteristic Catholic philosophical commitment to an objective moral order can be retained through the affirmation of a "core of rationality," a "principle of common humanity," and some consistent areas of "overlap" among distinct moral systems.[10] That to which access is given by the language of "natural law" is not a universal, culture-free realm of clear and logical moral propositions, but a "community of moral discourse" into which all can enter, bearing with them their special value-commitments. The phrase is used to indicate the processive and interdependent character of the quest for moral truth – even granted that truth does exist and can be known. What conversation partners bring to the dialogue is experience-based insight into moral obligation, which for Christians will include the experience of conversion and radical discipleship. Out of that experience, areas of basic moral concurrence are affirmed (respect human life and dignity), but values and obligations may take on a distinctive coloring (the radical present obligation to love even the enemy). As Bryan Hehir has stated, "the distinctive contribution of religious communities to the public debate lies in their religious values and moral insight." At the same time, effective public witness by these same communities depends on their ability to appeal to "those who do not share a single religious conviction."[11] Persuasive power within the community of moral discourse depends on the degree to which the language of each partner can resonate with the experience of others, as overlapping moral perceptions are clarified, refined, and redefined toward some consensus. The "natural law" language of Catholic ethics represents a commitment to this process. The renewal of the distinctive religious roots of Catholic ethics represents, on the other hand, a commitment to the truth of Christianity's special insights into the nature of moral relations, as expressed in the demand to identify with the needs of the other (even the enemy) as though they were one's own.

Boston College
Chestnut Hill, Massachusetts, U.S.A.

NOTES

[1] As Fuchs says, "Catholic medical ethics has generally presented itself as a philosophical ethics . . ." ([4], p. 1). Also, "no concrete ethics . . . can be developed out of faith," understood as a basic act of trust in God ([4], p. 5).

[2] Fuchs seems to adopt this solution in his present essay on pp. 5,6,8,14.

[3] Read the Sermon on the Mount in the context of *Matthew* 5–7; cf. J. R. Donahue, S. J., "The Parable of the Sheep and the Goats", [2].

[4] See the classic statement by J. Jeremias, [8].

[5] This favored phrase of Pope John was repeated in *Gaudium et spes* ("The Pastoral Constitution on the Church in the Modern World"), p. 4

[6] He refers to sexual expression as "the language of the body", for example in "The Transmission of Life", [10].

[7] A recent issue of *Health Progress* (July-Aug. 1986) includes an essay by Bernardin on

"The Consistent Ethic of Life" as applied to the issues of health care. The same issue includes the "Report of the Catholic Health Association Task Force on Health Care of the Poor", as well as several other essays in which the social dimensions of health care are addressed.

[8] See Thomas Aquinas, *Summa Theologiae* I–II .Q94.a4. This perspective is reflected by Fr. Fuch's remark that perhaps we today discuss problems of killing in "wholly other circumstances and possibilities" than those envisioned by the tradition [3], p. 90).

[9] This encyclical of Leo XII, titled in English, "The Condition of Labor", was issued in 1891. It began a papal tradition of addressing the particular questions of social justice that arise in the industrialized modern world. Key among these have been socialism, democracy, and capitalism as economic and social systems; the right to private property and the responsibilities of property owners; employment, labor, and labor unions; the duties and limits of government, and the proper sphere of subsidiary social bodies.

[10] G. Hughes, S. J.: "Is Ethics One or Many?" [7],p. 188. This way of posing the nature of a common human morality does not seem necessarily inconsistent with Aquinas' distinction of the basic and secondary principles of natural law in *Summa Theologica*, I–II, Q94, a.2, a.4.

[11] J. B. Hehir [5], pp. 211, also adds at the conclusion of his paper that "religious differences will yield different conceptions of how to address medical ethics and standards of justice" [5], p. 220). This does not preclude public debate in a pluralist society, but acknowledges significant influence in that debate of moral commitments that are religiously rooted.

BIBLIOGRAPHY

1. Collins, R. F.: 1986, *Christian Morality: Biblical Foundations*, University of Notre Dame Press, Notre Dame, Indiana.
2. Donahue, J. R., S. J.: 1986, 'The "Parable" of the Sheep and the Goats: A Challenge to Christian Ethics', *Theological Studies* **47**, 3–31.
3. Fuchs, J., S. J.: 1988, '"Catholic" Medical Moral Theology?', *Philosophy and Medicine* **34**, in this volume, pp. 83–92.
4. Fuchs, J., S. J.: 1965, *Natural Law: A Theological Investigation*, trans. H. Reckter, S. J. and J. A. Dowling, Sheed and Ward, New York.
5. Hehir, J. B.: 1988, 'Religious Pluralism and Social Policy', in this volume, pp. 205–222.
6. Houlden, J. L.: 1977, *Ethics and the New Testament*, Oxford University Press, New York.
7. Hughes, G., S. J.: 1988, 'Is Ethics One or Many?', *Philosophy and Medicine* **34**, in this volume pp. 173–196.
8. Jeremias, J.: 1963, *The Sermon on the Mount*, Fortress Press, Philadelphia.
9. *Optatam Totius* (Decree on Priestly Formation): 1966, Walter M. Abbott (ed.), *The Documents of Vatican II*, America Press, Association Press, New York.
10. Pope John Paul II: 1984, 'The Transmission of Life', *The Pope Speaks* **29**, 349–51.
11. St. Thomas Aquinas: 1964, *Summa Theologica*, Blackfriars, McGraw-Hill, New York.

KLAUS DEMMER

THEOLOGICAL ARGUMENT AND HERMENEUTICS IN BIOETHICS

INTRODUCTION

Sciences are in flux. They are subject to the growing complexity of their own questioning. It is with good reason that one speaks of open systems. New objects of knowledge are continually being taken up and integrated into the accepted system. But more may be said. The accepted presuppositions of a system are also subject to a critical analysis. These two phases of the scientific enterprise reciprocally condition each other. This has consequences on the precision of the formulation of the discipline's prevailing problems. Further, the methodology of each science participates in this process.

Moral theology is no exception to this process. Whenever new problems arise, moral theology critically reflects on the functional effectiveness of its methodological structure, and the plausibility as well as the consequences of its process of argumentation. The history of moral theology offers unquestionable support to this thesis. The purpose of the following reflection is to investigate this more closely. Such a point of reference can relieve something of the dramatic nature of the contemporary situation of bioethical debate. Parallel examples from its own tradition can be easily found also.

Each science attempts to remember its own past. The guiding question is whether or not fundamental paradigms have been exhausted in their ability to provide adequate explanations. Only when this has been determined is it possible to responsibly speak of a shift of paradigm. Because of its ambivalence, this phrase should be dealt with critically and cautiously. One can discover new paradigms; indeed, this happens when the chasm between empirical verification and theoretical explanation becomes disjointed. However, progress is also internal to the paradigm as it undergoes continual differentiations and extentions ([10]; [15]).

The moral theologian operates on many different fronts. On the one hand, the theologian must take into account the cognitive developments of the empirical human sciences. On the other hand, the moralist must not lose living contact with theology; the discipline within which he

103

Edmund D. Pellegrino et al. (eds.), Catholic Perspectives on Medical Morals, pp. 103–122.
© *1989 Kluwer Academic Publishers. Printed in the Netherlands.*

works is understood as strictly a theological science. It must be asked, then, whether one's moral reflection is in accord with general theological reflection; can it be the case that a moral problem is approached with inadequate theological tools? The developments in the fields of dogmatic and fundamental theology will also be relevant to the moral theologian; otherwise the moralist risks being pulled into a theological ghetto. This is particularly true for theological epistemology and the underlying theology of revelation.[1]

In the following reflections an attempt will be made to shed light on this problem. The theological argument serves as the point of departure. First, then, theology's validity and its limits in the context of bioethical dialogue is defined. Second, consideration is given in regard to a responsible approach to natural law arguments. Hence, there is the question of the hermeneutical mediation of faith and reason. Each concern can be illustrated with examples.

I. THE VALIDITY AND LIMITS OF THEOLOGICAL ARGUMENT

1. The Distinctiveness of Moral Truth

In order to gain a perspective on the significance of the theological argument, the distinctiveness of moral truth must be considered. Moral truth is, by its very nature, the truth of one's life project. It demands from the person a dedicated life commitment. The goals of one's life and the objectives of one's actions are outlined and presented through it. This sheds light on the peculiarity of moral reason. Moral reason is not primarily a predetermined and standardized form of reason; rather, it is dynamic in that it continually probes and discovers new possibilities. In this way, the goals and objectives for life and action are uncovered and evaluated in light of freedom. There is a dialectical mediation between theory and praxis.

Normative discourse and argumentation are set within this process, and it provides a guiding interest that cannot be eliminated. At the same time, however, it is held in flux because of reason's dynamism. It concerns itself, then, with justifiable plausibility or reasonableness. Concomitantly, it receives a growing recognition which is nurtured by meta-normative sources. It is a question of hermeneutical mediation. This mediation takes place between anthropological presuppositions and the perception of moral values[1]. What kind of correspondence exists between them? How can this correspondence be normatively grounded?

2. The Theological Argument

With what has been said, the ground is laid for the theological argument. What is required is an appropriately theological "depth-hermeneutic" through which the anthropological implications of faith can be revealed. The guiding question is this: what is the normatively relevant self-understanding which characterizes the believer as a believer? The answer to this question is of theological relevance to the moralist. This is the focus of his questioning and reflection as a theologian. Scientific discourse of God can only proceed from the context of human self-understanding.[2]

It should be clear, then, that there is no direct recourse to the normative "will of God". This has always been accepted in Catholic moral theology, which has never been representative of unreflective moral positivism. In like measure, the will of God has never been understood as an heteronomous authority in competition with responsible human self-determination. Rather, the will of God is always spoken of against the background of the classical teaching on the analogy of being. This teaching was and is the *articulus stantis et cadentis theologiae catholicae*. On this basis there is a highly complex theological and anthropological process of understanding at work. Nevertheless, an impression to the contrary occasionally arises because this process is often taken for granted[4].

3. The Insufficiency of Sacred Scripture

In Catholic moral theology, there is a general consensus about the insufficiency – both in terms of methodology and content – of Sacred Scripture. Scripture is no handbook of moral theology; nor does it present an ethical system. Whenever there is a question of normative ethics, it must be remembered that its assertions are bound to their original context. Scripture is not meant to be employed for all problems referring to the responsible formation of the inner-worldly areas of life. This becomes particularly relevant in those areas where human life is in question. Further, there is the fact that the complexity of the processes of ethical argumentation are not completely known in advance as they would be in a manual. To expect information in these areas would be to overstep the competency of Scripture[5].

This inevitable situation cannot be overcome by a hermeneutic of individual passages of Scripture – even if the perspective of the moral theologian should guide such hermeneutical attempts. One may not overlook the simple fact that the biblical author lacked the requisite awareness of contemporary problems. It is a fundamental principle of hermeneutics not to expect answers from a text which it never intended to provide. However, this does not imply any renunciation in principle of a scriptural argument; it only means that one must begin with a more comprehensive and more profound understanding of it if one wishes to use it. This means several things.

First, an individual text is not to be seen in isolation but inserted into the context of its pre- and subsequent history. Texts do not simply appear as bolts out of the blue. They are the result of a long process of insight and experience. The respective author stands within a tradition which is concretized in the text. In addition, texts exert a subsequent effective history. They are considered in new ways in changing historical contexts. That is, they are further developed in history. In this regard it becomes clear that texts are not to be read as isolated and well-defined sayings. It is much more important to discover their guiding concern. Hermeneutics is required to uncover the history of an insight; its task is to identify the tendencies of a text and simultaneously to further develop them. In this way, hermeneutical reflection becomes more than the reconstruction of the genesis of an insight. It becomes the creative process that further expands the insight. This perspective already expands the level on which hermeneutical reflection is effective. What is meant is that textual hermeneutics is accompanied by a corresponding "depth-hermeneutic". The implied self-understanding of the biblical author is to be brought to light. It forms an indispensable point of reference for any reasonable dealings with the arguments from Scripture. Here there is found an essential presupposition for the methodology of moral theology.

4. The Relevance of Faith

It is assumed that faith is normatively relevant. When one speaks of faith in this context, it is understood in the sense of the *fides quae*. That is, what is meant by faith includes the explicit declarations of faith[12].

Even here there is need of further clarifications. As before, a general consensus can be found within Catholic moral theology over the fact that direct behavioral norms cannot be deduced from the explicit declarations of faith. This would clearly exceed epistemological and methodological boundaries. The natural law tradition within Catholic moral theology eloquently attest to this fact. Nevertheless, one can speak of a maieutic function of faith in relation to moral reason. The subjective principle of cognition of the moral theologian is reason illuminated by faith. This is uncontested. However, the question is how this inspiration proceeds and what justifiable expectations one can place on it. This is the central problem of theological epistemology. With this in mind, what does the so-called maieutic function of faith mean? What is meant is that faith puts moral reason in the position to pose the proper questions. This is only possible when faith generates normatively relevant pre-understanding. It not only perceives the meaningfulness of the question, but at the same time, it gives at least an indication of the direction in which the answer is to be sought. If one does not work from this presupposition, then all talk of pre-understanding is irrelevant.

At this point it becomes evident that theological-moral questions exist in the context of the theology of revelation. In this area, Catholic moral theology recognizes the need for further work. What must be presented is a coherent concept of revelation in which the mediation of theological and anthropological categories constitutes the central concern. It makes no difference which model of revelation is used as a foundation: the scriptural understanding of "epiphany", the theoretically instructive model of the Middle Ages, or the contemporary model of self-communication[16]. Whichever model is used, there is one area in which they all converge; that is, they are all conventional atttempts to define a basic self-understanding. At this point this cannot be pursued further, yet one decisive consequence can be clearly indicated. Faith elicits reflection from moral reason; it inspires it. It does so in that it produces in the believer, through his understanding and interpretation of revelation, an understanding of himself and of the world. It is this understanding of one's self and of the world that is at the root of all ethical insights and expressions. The moral reason of the believer is bound to this self-understanding and understanding of the world. It does not float in free space. This may be expressed in another way: between the explicit statements of faith, anthropological implications, and ethical norms, there is recognized a relationship of correspondence and appropriateness. Because of this relationship the anthropological implications of faith have an undeniable mediatory function. This complex process is borne by understanding and interpretation. This is what is meant by the *intellectus fidei*. The theological argument in moral theology is based on and reflective of this process.

5. *The Anthropological Implications of Faith*

Without claiming to be complete, some examples can be briefly singled out. They illustrate the fact that the believer reflects on both the fact of revelation as well as its content. Yet the validity of these examples is not limited to the context of faith. One cannot see in them an exclusivity in terms of content. What is being underlined here is that these examples rest on a theological foundation; nothing more is meant, but also nothing less [2].

First and foremost is the undestroyable dignity of the human person. It is at the root of the principle of autonomy. It is based on the teaching of the creation of the human person in the image and likeness of God, and the new creation in Jesus Christ. Closely connected to this is the awareness of the singularity of one's existence as well as the unrepeatable nature of one's personal history. This first anthropological implication of faith reveals a perspective of the intimated uniqueness of the person over which no one has a claim.

Consequently, one is led to call to mind the fundamental equality and fraternity among all people. If God, through Jesus Christ, became human, then he makes all human beings equal to each other. Since God became a human being, human beings have become the measure against which all other things are measured (K. Barth). The collectivity of human nature is, however, in no way an adequate foundation for a further anthropologically legitimated and reflected equality. It is ambivalent. How it is interpreted, then, becomes the decisive factor in ethical discourse.

The quality of history is also a consideration. If God communicated himself to human beings in history, and if history is where man encounters God as the ultimate foundation of the meaning of his unique existence, then there can be no situation that can be excluded from the possibility of being given definitive meaning. This is reflected in a composed confidence in the face of extreme and burdensome situations.

Finally, the Easter event provides a key of understanding and interpretation in the task of making sense of one's death and of its historical anticipations. If death is not to be the definitive human and moral catastrophe, but rather a passage into eternal life, then there can be no historical situation which stands outside of this promise and its power to transform. For the Christian, death and life are, at their very roots, reconciled with one another. There is neither an escape from life nor an escape from death. The Christian grasps death at the culmination point of his life. This enables him to obtain his life[13].

These few examples show that the moral reason of the Christian operates within an anthropological system of coordinates. It is not presuppositionless. It remains the task of theological-moral reflection to dedicate more attention to these basic elements of theological anthropology. Hence, the teachings on justification and reconciliation must also be taken into consideration and their anthropological implications disclosed. The believer is freed from the paralyzing compulsion to seek self-justification in this world. He reconciles the immanent conflict of history. However, these thoughts cannot be pursued here. It is sufficient to have shown that between faith and moral reason there exists a complex process of mediation. Without this mediation, faith would remain extrinsic and would not be in a position to exert a clarifying historical efficacy on moral reason.

These thoughts can be pursued still further and determined in their moral theological relevance. The anthropological implications are neither formal nor lacking in content. However, they have also not been fully established with regard to their content. Rather, they occupy a middle position. Seen from this vantage point, they furnish a standard criterion. It is the task of autonomous moral reason to achieve the further determination of the content. In that, it is never sufficient to proceed only from individual fundamental anthropological elements. It is much more essential to hold in view the interconnected pattern of all anthropological implications. Only then will it become apparent that faith contains within itself an anthropological project. Yet this project is understood as open in terms of its content of meaning.

The required process of determination is not established *a priori*; rather, it remains embedded in the unforseeable interplay between all of the morally relevant historical circumstances. The anthropological proprium of christian ethics is the product of an open history.

6. The Character of Moral Knowledge

At this point a further presupposition must be mentioned which has not previously been indicated. It concerns the character of moral knowledge and is bound to the previous exposition on the character of moral truth. Initially, moral truth was characterized as the truth of a life project. This does not detract from truth's objectivity. It is discovered and simultaneously established by moral reason. The objectivity of its claim is bound to this overlap between discovery and establishment. When one recalls this relationship, then it is easier to understand the above-mentioned illuminating function of faith in regard to moral reason. Faith presents no developed teaching of morality, nor specific moral content. Were one to argue such a position, it would be based on a dual reductionism. First, it would rest in the underlying understanding of moral knowledge. Moral knowledge would be reduced to the reporting of a perfected pre-given objective moral truth. The creative function of discovery and determination would be ignored. Further, however, the fundamental model of revelation would not be free of a subtle reductionism. The theoretically instructive model would dominate. Revelation would be understood as the mere communication of statements – normative statements not excluded. In contrast to this, the consideration up to now should have made clear that the mediative function of the anthropological implications of faith gives rise to a creative and critical potential for knowledge and insight which in turn is explained in an historical process. The anthropological implications provide guideposts; they serve as directional signals.

This provides the opportunity to turn to another form of reductionism that is often present. It does not pertain directly to the normative discussion, but belongs most originally on the level of a theory of action. It has no little significance for normative ethics. What is to be considered here is the co-ordinating relationship between motivation and intention. Occasionally, the position is maintained that faith provides the motivation for action; that it operates, originally, on the level of the *goodness* of the action; on the other hand, the *rightness* of the action is delegated to autonomous moral reason ([4], p. 54 ff.). This position is not wrong; however, it has the impression of being an oversimplification. It is not sufficiently recognized that motivation and intention are mutually and inextricably bound to one another. They share a common point of reference, namely, the coordination system of theological anthropology. Certainly it lies within the competence of autonomous moral reason to explicate and determine the anthropological implications of faith in their immediate relevance for praxis. But here what is obviously dealt with is a relational autonomy. It is linked to a pre-understanding that arises out of the insight of faith. Moral reason, so understood, develops a christian context of intention which consequently enters into the individual intention of the person acting. Life goals and behavioral goals are produced through this intentional context. A judgement of rightness is established, which refers back to the presuppositions of theological anthropology. Only against this background is it possible to understand the function of motivation. Motivations can only have an impact on the one who acts, i.e. on one's personal goodness, if they are met with the corresponding correct (right) content. Rightness and goodness refer to one another. The motivation arising from faith seeks intentions that originate from moral reason illuminated by faith. Similarly, intentions require corresponding motivations in order that they might be brought into reality.

7. The Hermeneutic of Magisterial Documents

If one reflects on the theological argument as a whole, then the intervention of the Church's teaching office in questions of morality appears in a new and often in a different light than is usually the case. In papal teaching explicit recourse is often taken in reference to human dignity and to a christian image of the person[8]. Of course this is not wrong. But an uncritical and short-handed way of dealing with such concepts is to be avoided as much as possible. The question remains, how is human dignity and the christian image of the person to be concretely understood? Hermeneutical reflection is necessary. The impression of moral positivism is to be avoided. Perhaps the presumed correspondence between faith and the moral consequences drawn from it are not as apparent as they might first appear. Is there at all something like an ethical pluralism within the Church's *communio*, which is conditioned by the complexity of the previously described process of mediation ([6], p. 43)?

What is needed is a clearly developed hermeneutic of magisterial documents. What this means is not just an indication of the level of authority of the documents; this is usually made clear from the language and the character of the document itself. For example, an encyclical has

a much higher level of authority than either a "*Motu proprio*" or an
"*exhortatio apostolica*", to say nothing of the numerous papal dis-
courses given at various occasions. Rather, what is of importance are
the various cognitive presuppositions that enter into a teaching document and determine its content. What philosophical presuppostions are
present? Why are they being put forward? Do they take account of the
complexity of the moral problem to be solved, or are they insufficient to
meet this demand? Ultimately, the moral theologian is not alone in
bearing the responsibility for his philosophical tools. This is a requirement for all of theology. Yet, this is not achieved with this first postu-
late. Further points of view must be brought forward. A hermeneutic of
magisterial documents requires that the ever newly determined tension
between reasonable arguments and the respective doctrine be carried on
in all seriousness. Certainly one can maintain a correct teaching without
totally convincing arguments[14]. But if the teaching is correct, then, in
principle, it must be made clear through reasonable and plausible
argument. Furthermore, it must be considered whether the Church's
teaching office uses fundamental concepts of philosophical anthropology
consistently. This is to be considered, for instance, in regard to the use
of the key concepts "person" and "nature". Particularly in the field of
bioethics, such a hermeneutic is essential because this discussion involves concepts lying at the very point of intersection between the
empirical human sciences and philosophy. Does theological reflection
perhaps utilize philosophical concepts that no longer correspond to the
state of scientific research? It is clear that the hermeneutic of the
teaching documents of the Church involves a very difficult task. This is
true regardless of the question of the historical context from which such
documents arose, though even this question is also helpful in providing
information regarding the concerns and meaning of the documents of
the Church's magisterium.

Summarizing what has been said, it is clear that theological argument
has an essential place in moral theology. When such theological argument is used, the boundaries of that which can be reasonably argued for
and the complexity of the required process of mediation must be kept in
mind. When one fails to do so, one without fail, comes to a dangerous
reductionism in methodology, and consequently to distortions of the
truth.

II. THE RESPONSIBLE USE OF NATURAL LAW ARGUMENTS

1. The Relationship between Theology and the Natural Sciences

The moral theologian sees creation as a good, as the creation of God over which man is a responsible steward. Man does not stand at a distance from nature; he participates in it. As a creature of God he is part of it and conditioned by it. However, man also has the responsibility to fashion it.

Such a theological qualification does not mean that the results of the natural sciences and theological statements can be brought immediately into relation with one another. Between the two a highly complex process of mediation takes place. This process needs to be briefly described.

It must be realized that between these two disciplines there does not exist a complete antithesis. To be sure, the results of natural scientific research must be reflected on philosophically and theologically; this is obvious to both philosophers and theologians. Conversely, the natural scientist must take into consideration his own tacit philosophical and theological presuppositions. There is no presuppositionless science. Every science is established in an historical range of experience.

When this is considered, the role of the theologian can be rather easily determined. The theologian seeks to incorporate sound results of scientific research into his horizon of understanding and interpretation. Only when these results are understood and interpreted are they incorporated into theological reflection ([9], p. 88ff). Consequently, only as understood and interpreted are they incorporated into ethical discourse. In such a way, the theologian is assured that there can never be a contradiction between faith and science. Yet one may well exist between the way the natural sciences and theology formulate their respective theories. In order to be clear about this, the status of scientific and theological theories needs to be considered. How are the respective theories established? The moral theologian can think that behavioral norms are nothing other than established theological-moral theories. However, this is not all. The theologian must also guard against the ever present danger of an over-interpretation of the facts. What can be justifiably inferred from the facts? These are generally accepted postulates of any interdisciplinary dialogue.

2. The Theologian as a Passive Partner in Dialogue

In the first instance, the theologian assumes the role of one who learns from the natural scientist. After all, one's image of God is reflective of one's understanding of creation. Hence the results of the natural sciences are always cause for reflection on the part of the theologian. This is uncontested. What must be determined is "how" this is done. In this regard the work of Jean Ladrière is helpful[11]. In his view the results of the natural sciences are not comparable to pure informational material that could be readily included in any number of structures of thought. Rather, they alter the self-understanding of the subject. This does not happen directly and immediately, as might easily be thought, but rather in an indirect and mediate way. The change is set within a long historical process. In short, the presuppositions of theological reflection which are open to the natural sciences undergo a slow yet continuous historical transformation. The obvious effects are repercussions on theological reflection.

Here a challenge to the moral theologian can be conceived which was hinted at earlier. Is his understanding of normative nature based on paradigms that are no longer suited to the state of natural scientific research? The Catholic moral theologian does not support any kind of extrinsic moral positivism; nor is he confronted with the binding will of God in those situations where the natural law argument in its complexity experiences its limits. God is not an expedient stop-gap in the face of the historical risks in the formulation of ethical theories. Just as little is God conceived as a rival to the onus man freely assumes by virtue of autonomous moral reason. Man cannot use faith to avoid the responsibility demanded of him on account of his intellect.

However, faith supplies man with an interpretive horizon in the sense this was described above. The theologian must try to take into consideration the anthropological consequences of established developments in natural scientific research. Do these developments contain dangers to the underlying conception of a comprehensively successful good life? Or do they hold the possibilities of a higher quality of life?

3. The Development of a Paradigm: The Principle of Totality

The history of moral theology supplies examples of such changes. Theology had adapted to the progress made in the natural sciences and has changed the criterion for the formation of ethical theories. The development of the principle of totality serves as an example[7]. The question can be left open whether there has been an actual shift of paradigm or whether the accepted paradigm has only undergone further differentiation. In any case, it cannot be denied that there has been a discontinuous expansion. As is well known, the principle was initially reduced to a physicalistic and individualistic understanding. The dominant axiom was simply *pars propter totum*. A therapeutic operation for a diseased organ or bodily function was considered permissible when no other possibility existed to secure the well-being of the organism. Moreover, this required a correspondence – which was strictly interpreted – between the employed means and the end they attempted to reach. Both had to move on the same level; that is, a bodily illness was answered by a corporeal intervention.

This changed over time, and some of the stages of this discontinuous development can be briefly mentioned. The impetus for the development was given by Pius XII. He permitted a medical intervention on a healthy organ for preventive or even therapeutic reasons. An initial expansion of the principle, then, was given. However, moral theological reflection had already begun to find a more flexible relationship between the means employed and the end intended. This was necessitated by the discovery of new medical and technical possibilities, e.g., neuro-surgery and organ transplants. The intended end could no longer be understood in individualistic and physicalistic categories. It lay now on the level of interpersonal and personal goods. However, it must be added that any expansion of the principle in these terms was only possible because medical experience was able to show that expected dangers were no longer present. The traditionally strict interpretation of the principle of totality was based on and reflective of the available medical and technical possibilities. Its central focus was always the protection of the person. But now the tradition is not as limited and rigid as it might have first appeared. The beginnings of an expanded understanding of the principle of totality appears already in the classical manuals of moral theology. This is seen in their allowance of the sacrifice of one's own bodily integrity when it was the only means of securing freedom. One recalls the often cited example of the chained prisoner who cuts off his hand in order to gain him freedom which cannot be attained in any other way ([17], p. 265). With this historical background one should not be surprised when in the recent past the principle has been further developed. It parallels the expansion of the concept of therapy. This is used, for instance, in the treatment of a phobic fear of pregnancy through the administration of contraceptives. The good end and the employed means, however, do not lie on the same level. In another instance one can refer to a hysterectomy in the case of a damaged uterus with the end making possible an anxiety-free conjugal life. In order to understand these examples, reference to the underlying anthropology of sexuality is required. In the background stands the altered view concerning the openness to offspring of the individual marital acts.

4. The Anthropological Consequences

This process, which was set in motion through medical and technical advancements, allows a better understanding of the relevant philosophical and anthropological categories. They became clarified as one advanced further into their content of meaning. The fundamental understanding of the person serves as an example. In the wake of the extension of the principle of totality, an understanding of person was achieved which combined the aspect of "substance" with one of "relation". The traditional teaching on substance was enlarged through the aspect of relation. Naturally, this process is not concerned with something absolutely new. It had been recognized earlier that the "person" is substance-in-relation. However, now the emphasis has shifted and is less ambiguous. It is necessary now for the moral theologian to strike a balance between both aspects. More exactly, a new balance that takes into account both technical advancements as well as anthropological presuppositions.

It is clear, then, that the concept of person experiences a change of meaning as a central metaphysical category of theological moral argument. The same is true for the relationship between person and nature. Man's possibilities of intervention in his nature are able to be set in a wider context because of a more flexible understanding of these two fundamental categories. This is on account of a reciprocity which exists between the questions formed from natural scientific and philosophical-anthropological concerns. Standing alone, neither has a self-sufficient existence; rather, they stand in an overlapping and reciprocally conditioning relationship to one another. It is necessary to look at this overlapping relationship more closely.

5. The Theologian as an Active Partner in Dialogue

Up to now the focus has been on the moral theologian in the role of one who learns. However, the critical use of the natural law argument in moral theology requires that the focus can be set equally in the opposite direction. The moral theologian is not only one who learns but also one who gives – one who contributes to the discussion. He is required not only to reflect critically on the results of the empirical sciences, he also causes the natural scientist to reflect. He makes an independent and active contribution to the interdisciplinary dialogue.

In order to clarify this contribution, a brief reflection on the characteristics of the natural law argument as well as the underlying understanding of normative human nature is required.

Man has an anthropologically founded project in mind when he approaches empirical nature. This project is not sufficiently derived from nature alone. It transcends empirical nature. The facticity of human nature is not normative. Only human nature which is understood and interpreted in light of such an anthropological project merits such a qualification.

This is relevant for the contact the moral theologian has with the natural scientist when the natural law argument is under discussion. The moral theologian may not succumb to the dangers of naturalism. It cannot be denied that the facticity of nature gives indications that must be considered and which provoke one to a position from which to understand and interpret them. But again, nature is in no way normative in its facticity. If one does not take this fact into consideration, one risks committing the naturalistic fallacy.[3] It cannot be emphasized enough that normative nature is established through the overlapping relationship existing between its being understood and its being interpreted. Certainly there is the initial phase in which one of necessity learns from nature. But the phase involving its understanding and interpretation cannot be forgotten. This is done in light of meta-empirical criteria.

The same point can be made in another way. In relation to his nature, man is faced with a responsibility to fashion and shape it. This begins already in his understanding and interpretation of it. An anthropological pre-understanding not only determines the problem; it at the same time directs the process of reflection and points the direction in which solutions may be sought. The facticity of nature is made transparent through understanding and interpretation. Hence, the same phenomenon can be understood and interpreted differently because of different underlying presuppositions. For the moral theologian this means that his understanding of normative nature has to pass through the filter of a theological anthropology. If not, he is guilty of an ontological and methodological reductionism.

6. The Finality of Nature

An example of this is seen in the frequently used reference to the finality of nature. What is its relevance for ethical arguments? A brief look at the recent past can help to answer this question adequately. In the reaction to *Humanae Vitae* the accent was clearly on a personalistic understanding of human nature. The nature of the person as distinguished by his intellect was meant to be normative, ([6], 15). Theologians strove to avoid any suspicion of a naturalistic fallacy. It could be debated whether *Humanae Vitae* actually commits such an error or whether it is the case of approaching empirical nature with a particular concept of marital sexuality. In the present context those considerations lie too far afield. It is noteworthy, however, to notice that the pendulum seems to be swinging back today. That is, the outline of a neo-naturalism is sketched. This is not only done on the part of the natural scientist, but on the part of moral theologians and ethicists as well. This problem can be briefly considered.

It consists in the tendency to tacitly anthropomorphize nature. Theologians speak of a finality of nature and then argue that man is under no obligation to be more natural than nature itself. Man becomes an apprentice to nature. Nature gives him its clues and man meaningfully imitates them. What this means, of course, hinges on the understanding of "meaningful". The meanings of terms are determined by particular epistemological and anthropological presuppositions. Reference to an often discussed example can clarify this. One speaks of the "principle of abundance" in nature. Nature handles human life extravagantly. This is seen in reference to the high number of spontaneous abortions. Where, then, does the above-mentioned naturalism lie? It is present when one tacitly assumes a eugenic finality of nature; that is, by putting forward the thesis that nature eliminates badly damaged life. The naturalistic conclusion drawn from such an assumption can be formulated in the following way: when the natural processes pass to man's own responsibility, i.e., when it is a question of man's right of disposing of human life, then the same criteria hold as for nature. His possibilities of disposing of human life are determined through the principle of abundance. When nature fails in the pursuance of its eugenic finality, then man would be allowed to come to nature's aid and attain the finality it is unable to.

However, serious questions are raised in face of this position. Is it question of the basis of a meaningful imitation of nature, or is there a crude naturalism at issue? Natural processes like that of spontaneous abortions are certainly cause for reflection. However, these instances supply neither immediate nor practical criterion for responsibly dealing with human life. Through the inspiration of anthropological archetypes, autonomous moral reasoning has to develop criteria through an independent process of reflection. The question, then, is: on the basis of these anthropological presuppositions what is meant by meaningful life and survival? The natural processes are interpreted in view of such pre-understandings. With this perspective, man projects anthropological finalities into nature. They supply the criterion for the disposing of human life.

However, the theological dimension of this problem is also to be considered. How, for instance, is the "principle of abundance" to be explained when nature is understood as the creation of God? This deserves a closer examination. First, it is necessary to ask if the expression "principle of abundance" is used correctly? Nature does not handle human life extravagantly. Any embryos which are lost through a spontaneous abortion suffer from such gross chromosomal aberrations that they have no capability of surviving. They are not eliminated in the sense of a eugenic finality; they are incapable of survival. The entire process would be at best an inquiry into a naive understanding of God's good creation.

The problem of God returns here in a modern form. God has not created the best of all possible worlds. Human life is endangered at all stages of development and in all situations of its existence. However, the justification of an artificially provoked risk to human life cannot be deduced from this fact. This would be an obvious over-interpretation of the facts. On the other hand, however, when it is said than man applies or projects into nature an anthropologically founded schema, the central concern is to moderate the risks immanent in nature. Man is under the obligation to humanize and to refine his nature. Within this perspective lies his anthropologically centered understanding of God as creator. It may seem one-sided to demand that man correct his conception of creation in view of the new knowledge of molecular biology. Such a demand could be taken into account only if at the same time man's relationship vis-à-vis creation were altered. Both aspect belong together because creation cannot be understood exclusively through cosmological categories; human understanding, interpretation, and formation of the world and its natural processes fall under God's good creation as well.

These brief remarks are sufficient. They are of necessity incomplete. They must be completed with a reflection on the personal status of the embryo in its earliest stages. Yet even then, the attempt must be made to find concrete criterion for dealing with human life. This is an area that lies outside the scope of this paper. It is enough to have emphasized that the processes of nature do not contain in themselves direct practical criteria for moral action.

7. The Understanding of Pure Chance

Another example that can be considered is the uncritical use of the term "pure chance". It is not used in same way by the natural and human scientist. The word is always used when dealing with the so-called experiments within nature, but it is not always clear what concretely is to be understood by the term in this context. Does it refer to the absence of causal laws in the sense of classical mechanical physics? In this sense it would not be of serious interest to the philosopher or theologian. What had been advocated for a long time in the field of physics was applied only to the area of molecular biology. The absence of causal laws does not mean absolute chaos. It only means that the interaction of laws and facts is subject to a relationship that is indeterminate. Because it is clear that not all laws are known, the meaning of the term "pure chance" must be accurately described.

The situation would be completely different if the term "pure chance" was understood as the absence of the philosophical principle of causality. One would disavow, in this case, the metaphysically necessary relationship between cause and effect. When this takes place, however, the natural scientist oversteps his competency. He makes a jump from the natural scientific to the metaphysical level and draws immediate theological consequences, i.e., the denial of creation. In such an instance, the natural scientist is open to the criticisms of the philosopher and theologian because his conclusions are not discovered through the research results of the natural sciences, but rather originate from a tacit philosophical or even ideological pre-decision.

It remains to be said that pure chance does not describe an immediately operable ethical category. There is a strict parallel with the term "risk". It is up to the human mind to project a structure of organization onto nature. Consequently, in his dealings with nature it falls to his responsibility to exclude pure chance or at least to bring it under control. Again, in order to do so, anthropological criteria are required. Only under this presupposition can an understanding of pure chance be considered relevant in the area of the formation of ethical theories.

Here the moral theologian faces epistemological problems which often times are insufficiently recognized in moral theology.

CONCLUDING REMARKS

It was asked at the beginning of these reflections whether or not moral theology has something to learn from the general epistemological discussion. This question can only be answered with a resounding yes. However, the models of the natural sciences cannot be applied too quickly to the human sciences. In addition, the autonomy of moral truth must be taken into consideration. The formation of ethical theories follows its own proper laws.

One should deal critically with the term "shift of paradigm"; it too easily dissolves into a fashionable term. In fact, scientific developments and differentiations usually develop within the assumed paradigm. They make up what is called normal science. Shifts of paradigm do not happen abruptly; on the contrary, they are prepared for step-by-step through the development of normal science (P. Feyerabend).

Moral theology stands in a balanced, fluctuating, and overlapping relationship with the empirical human sciences. The moral theologian is both the apprentice and the teacher. On the one hand, he must take into consideration the morally relevant results of scientific research and integrate them into his conception of normative nature. On the other hand, he must, in the name of his own autonomous way of questioning as an ethicist and theologian, always voice criticism when the natural scientist exceeds the boundaries of his own professional competency. He may not submit to the hegemony of the factual sciences in terms of either their method or content.

Pontifical Gregorian University
Rome, Italy

NOTES

[1] See the theological development that has taken place after Vatican II, especially in the field of fundamental theology.
[2] This aspect has been focused on especially in the transcendental theology of K. Rahner.
[3] See the discussion following *Humanae Vitae*.

BIBLIOGRAPHY

1. Demmer, K.: 1985, *Deuten und Handeln: Grundlagen und Grundfragen der Fundamentalmoral*, Herder, Freiburg im Breisgau.
2. Demmer, K.: 1973, 'Moralische Norm und theologische Anthropologie', *Gregorianum* **54**, 263–305
3. Demmer, K.: 1979, *Sittlich Handeln aus Verstehen: Strukturen hermeneutisch orientierter Fundamentalmoral*, Düsseldorf, Patmos-Verlag.
4. Fuchs, J.: 1982, *Christian Ethics in a Secular Arena*, Georgetown University Press, Washington D.C.
5. Fuchs, J.: 1986, 'Christian Morality: Biblical Orientation and Human Evaluation', *Gregorianum* **67**, 745–783.
6. *Gaudium et Spes*: 1962, in W. Abott, S. J. (ed.), *The Documents of Vatican II*, Geoffrey-Chapman, London.
7. Hamelin, A. M.: 1966, 'Das Prinzip vom Ganzen and seinen Teilen und die freie Verfügung des Menschen über sich selbst', *Concilium* **2**, 362–368
8. John Paul II, *Redemptor hominis*.
9. Korff, W.: 1978, 'Wege empirischer Argumentation', in A. Hertz (ed.), *Handbuch der christlichen Ethik I*, Herder, Freiburg pp. 83–104.
10. Kuhn, T.: 1970, *The Structure of Scientific Revolutions*, University of Chicago Press, Chicago.
11. Ladrière, J.: 1972, *La Science, le monde et la Foi*, Casterman, Tournai.
12. MacNamara, V.: 1985, *Faith and Ethics: Recent Roman Catholicism*, Georgetown University Press, Washington, D.C.
13. *Philippians* **1**, 19-26.
14. Pius XII: 1950, *Humani Generis*.
15. Rorty, R.: 1979, *Philosophy and the Mirror of Nature*, Princeton University Press, Princeton, Ch. 1–4.
16. Rotter, H.: 1986, 'Christlicher Glaube und geschlechtliche Beziehung', in K. Golser, (ed.), *Christlicher Glaube und Moral*, Innsbruck-Wien, pp. 19–37.
17. Zalba, M.: 1953, *Theologiae Moralis Summa II*, La Editorial Catolica, Matriti.

MONIKA K. HELLWIG

THE DOCTRINAL STARTING POINTS FOR THEOLOGY AND HERMENEUTICS IN BIOETHICS: A RESPONSE TO KLAUS DEMMER

The following remarks are made from the perspective of a systematic theologian and will therefore focus more particularly on those fundamental questions for moral theologians which are closely related to changing perspectives in systematic theology. It is, in any case, in this area that questions over the role of the hierarchic *magisterium* in moral issues, questions about the scientific information factor, and questions about the mediation between these two factors must all be considered and resolved.

The paper presented by Dr. Demmer raises several key questions: first, the impact on the process and presuppositions of moral theology of cultural and social change and of scientific and technological advances; second, the relationship of the findings of moral theology to revelation, as formulated in Scripture and tradition, and as found in the whole of life; third, the dependence of presuppositions and arguments in moral theology on systematic theology; and finally, in the light of all of the above, the role of the hierarchic *magisterium* in the process of moral theology.

As to the first of these issues, it would seem that Professor Demmer has identified the correct point of departure. No matter what theological positions one may hold, one must operate within the historical and social realities of the real world in which knowledge and understanding in any intellectual discipline are necessarily in flux. Most theologians and believers will admit this readily in principle, but, as Dr. Demmer very discreetly hints, there is not total agreement on what this means in practice. There are certainly some who would like to limit the change to a process of adding on new applications of old principles and fitting new data into the old theoretical framework, This simply does not fit either the systematic or the practical exigence of the discipline. There comes a point from time to time when new questions or new data invalidate the old theoretical framework and the latter must be questioned and re-shaped.

Clearly, resistance to this may be due in some measure to a general state of human intellectual lethargy, but in the case of moral theology it

123

Edmund D. Pellegrino et al. (eds.), Catholic Perspectives on Medical Morals, pp. 123–128.
© *1989 Kluwer Academic Publishers. Printed in the Netherlands.*

is also related to the second issue raised, namely, the understanding of revelation. What we mean by revelation and how we think it happens is foundational in all aspects of religious life and teaching and therefore tends to define the process of theology and the role of the theologian. Fundamentalists, whether of the Protestant biblical type or of the Catholic papal-magisterial type, expect to read the answer to moral questions directly out of a documentary record of revelation. The process of moral theology (as of doctrinal or dogmatic theology) is seen as one of deduction from principles laid down in a documentary deposit of revelation. Revelation itself tends to be understood here as consisting of propositional formulae established and recorded in the past, and now enshrined and preserved by a clearly defined authoritative source – either the text of the Bible or the hierarchic magisterium. Most theologians of our times have rejected any such understanding of revelation. It is invalidated by its inadequacy to embrace the real and urgent questions which human persons must solve in order to live and form communities. Indeed, bioethics amply illustrates that inadequacy; answers to new and complex questions about life, health and resources cannot be drawn by deduction from any principles contained in a deposit of revelation. Even when we allow that the above is a "pure type" and that few church people actually think in such simplistic terms, it must be said that a static, propositional model of the content of revelation is simply inadequate to ground the process of Christian moral theology.

Struggling for a theology that is viable to cope with the questions that life really sets us, we have in fact come to an understanding of revelation that is in some sense a return to biblical ways of thought. It is a dialogic model in which God is at all times and all places self-revealing, through nature, conscience, human relationships and experience, history in general, and the religious traditions in particular, and in which human response is at all times co-constitutive of the dialogue, shaping its imagery, analogies, language, and codification. At the heart of this dialogue Christians discern the person of Jesus, but this does not terminate the process at a particular point in history, nor does it dispense Christians from their own engagement in the continuing dialogue. Such continuing creative engagement involves respectful attention not only to the past and the formulations emerging from it in one's own tradition, but also to the present and to experience and understanding arising outside one's own tradition.

Such a view suggests an attitude to the findings of science and the changing patterns of culture and society which does not make a sharp distinction between revelation and reason, but rather subsumes all observation and reflection into a perspective of reverent attention grounded in a posture of gratitude for existence and focused towards communion with the Ultimate, realized in some measure through deepening non-exclusive human community. Such an attitude and such an understanding are entirely coherent with the traditional Catholic sense of the continuity of faith and reason and of the interpretation of nature and grace. It suggests a wholly consistent contemporary approach to the traditional grounding of moral theology in "natural law", where natural law is not some static depository of ready-made answers to all questions, but rather a process of observation, dialogue, reflection, and argumentation.

This connects directly with the third issue raised by the intervention of Dr. Demmer, namely, the question of dependence of presuppositions and arguments in moral theology on systematic theology. If systematic theology is seen in the mode of "dogmatic theology", that is, in terms of an essentially finished body of doctrine needing but to be explained and passed on while remaining always simply the same, then Christian moral theology is likely to have established a path of its own. There will then have been an initial dependence on systematic theology in terms of the doctrines of creation, revelation, sin, redemption, allowing within that frame of reference for the grounding of moral theology in the "natural law", that is, in reason exercised within well-defined and unchanging boundaries. In this perception of the discussion of the belief system as "dogmatic theology" there is no contemporary transformation to trouble the moral theologian, nor yet to offer new avenues of approach to the new problems raised by technology, science, and changing culture.

However, if doctrinal discussion is seen rather in the mode of "systematic theology", then it undergoes some very radical transformations as the questions about belief are asked from the viewpoint of different societal and cultural situations, in the language of different philosophical systems and influences, and perhaps by wider and more varied circles of participants. In this case there will surely be both tension and mutual benefit in the relationship of doctrinal and moral theology. The dependence will be to some extent an interdependence inasmuch as systematic theology, at least in our times, comes to be understood more and more as a reflection on the praxis of living the Christian life.

In academic circles systematic theology has become progressively more humble in its epistemology; we claim less and less for knowledge and more and more for creative projection or conjecture or analogical construction. We acknowledge more and more clearly that all theology is anthropology projected beyond the limits of experience, while at the same time we maintain that this does not make the project untrue or useless, but means that we admit frankly that we are dealing with reflections or dim mirror images of the ultimate in the contingent. But precisely because we acknowledge theology as fundamentally anthropology projected, and because we are aware that we are reflecting on our praxis to arrive at an understanding, we also have more grounds for dialogue with moral theologians about contemporary and perennial issues in human conduct and decisions.

We refer back to the ancient themes of Christian belief but we do not see them as static or alien. We appeal to revelation but in the realization that God is at all times self-revealing in human experience and that the formulation of revelation in words is a continuing human endeavor. We appeal to creation as ground for a posture of gratitude and responsibility, knowing, however, that our reflections and any practical conclusions drawn from them must be guided not simply by facticity but by finality in a world in process of becoming. We refer back to the theme of original sin and of redemption from the heritage of sin and suffering, but we see sin not as an obscure past event casting a shadow over the individual's relationship with God, but rather as a pervasive disorientation due to the cumulative effect of destructive deeds – a disorientation affecting not only individuals, but the relationships between them, the social structures in which those relationships are expressed, and the systems that hold the social structures together in our world. Likewise, we see redemption not as a matter of inscrutable heavenly book-keeping nor yet as a purely inward, immaterial transformation within the human consciousness, but as the restoration of God's reign throughout creation through the transformation of human society in all its aspects, relationships, structures, systems, and activities.

In such a context we see Jesus Christ as accessible to us in his humanity, observable in his historical life as far as that has been recorded, and knowable particularly in his impact on history through his followers, not excluding the present generation. In other words, Christology and soteriology are not simply codified and unchanging explanations of a past and unchanging event, but rather a continuing dialogue between Jesus Christ as he is progressively and successively known and his followers as they attempt to live by the vision that he inspires in their changing circumstances and their expanding power in the world and in the universe. Church, then, as the community of the disciples of Jesus through history, becomes more and more a pliable and indefinitely adaptable instrument whose shape and character are determined not simply by the events and decisions of the past but much more by the finality to which it is intended, namely, the restoring of God's rule in all aspects of human life and all phases of creation – a restoration whose mode and process we must discern as best we can by the reflection on our praxis, which is centered on the praxis of Jesus himself.

This leads to the final question raised by Dr. Demmer's paper, which has to do with the role of the hierarchic *magisterium* in the process of moral theology. It is an open secret that the understanding of the process and substance of systematic theology as sketched above is not widely shared among those who staff the structures through which the Roman *magisterium* is exercised. It is, therefore, not a matter for surprise that the period since the Second Vatican Council has been marked by as much if not more tension and conflict than the time before. Nor is it surprising that the matter should focus on issues in moral theology having to do with human life, sexuality, and family. In a more static view of revelation, doctrine, theology, and church, accompanied by a more static view of nature and of natural law, the rules for human behavior in the private sphere cannot change because there cannot be any grounds on which such a change could conceivably be justified. Such grounds may arise in public issues when the shape of society and technology change, but in private issues the changing social structures and technology are apt to be seen as irrelevant because it is axiomatic that human nature does not change. Moreover, because there can be no change other than the adding of new applications by extrapolation, it appears not unreasonable from this point of view that the *magisterium* should have been progressively centralized and concentrated into a single staff in one place and with very restricted access to experience – wholly male, wholly celibate, all with an almost identical training, drawn together in a rather isolated social life, tending to be more in dialogue with one another than with the rest of the human community.

It is clear that the role of the *magisterium* in the process of theological, and more especially moral theological, reflection and formulation calls for urgent and extensive consideration if one approaches the matter from the viewpoint of more dynamic contemporary perceptions of reality and of human life within it – the viewpoint that more generally characterizes the theology of the universities in our times. We need to examine historically what happened to the process of consulting the faithful in matters of doctrine in the shaping of magisterial teaching. We need to ask how the *magisterium* came to shift from the collegiality of the churches as represented by their episcopal leaders to the unicity and particularity of a single see in the historical development of the life of the community of disciples through the ages. It is in the light of these historical inquiries that we need to consider the interaction of theologians as the theoretical experts, hierarchic leaders as the pastoral experts, and the faithful as the practical, experiential experts. If the task of the moral theologian in relation to the hierarchic *magisterium* is highly problematic in our times, it is because these questions have not been satisfactorily resolved, and because the apparent solutions unsuccessfully being imposed are invalidated by their inability to respond to the real problems of real people.

Georgetown University
Washington, D.C., U.S.A.

JOHN COLLINS HARVEY

A BRIEF HISTORY OF MEDICAL ETHICS
FROM THE ROMAN CATHOLIC PERSPECTIVE:
COMMENTS ON THE ESSAYS OF
FUCHS, DEMMER, CAHILL AND HELLWIG

The magnificent advances in medical science and praxis since World War II bring problems concerning moral behavior that should be addressed by all thinking individuals. Molecular biology has discovered so much about the structure and function of the individual cell that one could compare the cell now to the galaxy in which we live and the molecular biologist to the astronomer who has trained his telescope on the far distant stars and revolutionized our conception of the cosmos. In like manner, the molecular biologist has unravelled many of the secrets of the cell.

We know much about DNA, the fundamental basis of life. We have discovered the genetic code and we have learned how to cut and splice DNA so that we can manipulate and modify genes and their products. We can modify the somatic forms of cells and, indeed produce new forms of cells. We can influence the genetic composition of these new forms of cells and thus bring about new evolutionary forms of life at the microcellular level. In the future this surely will be possible at the animal and human levels.

In medical praxis marvelous technical advances have brought about the development of machines that may substitute for the function of organs that are so diseased they are non-functional. We have learned the techniques of transplanting organs and overcoming the immunological incompatibilities ensuing from the strange tissues that cause rejection. Thus, we can intervene to maintain the functioning of animals and man in a physiological way when in the past the animal or man would have died. We have conquered many infectious diseases that in the past wiped out large segments of the population or created great morbidity with altered functional capacity of individuals leading to their total dependency on others. Poliomyelitis has been eliminated from most of the advanced westernized countries of the world. Indeed, smallpox has been eliminated from the world! Yet, with such great advance in control over infectious disease we still are faced with the evolution of new forms of infectious agents, particularly viruses, which bring new epidemics

Edmund D. Pellegrino et al. (eds.), Catholic Perspectives on Medical Morals, pp. 129–144.
© *1989 Kluwer Academic Publishers. Printed in the Netherlands.*

such as the retro–virus HTLV-3, the cause of Acquired Immune Deficiency Disease. We remain helpless in the face of such developments for a while, until the discovery of a cure occurs. During this interim we can only use epidemiological measures in the hope of preventing the spread of the disease.

We have learned to overcome certain forms of sterility by harvesting spousal eggs and sperm and combining them in the testtube and reimplanting the developing embryos into the womb of the wife. Indeed, the methodology is even being used outside of marriage by some practitioners for sociological, psychological, or physiological convenience.

Medical praxis has played a very important role in the explosive growth of the world's population. We have developed techniques to promote fertility. We have by control of infectious diseases reduced mortality in infancy and childhood and, thus, increased population growth. This growth has also been enhanced by developments in praxis that alter the morbidity and mortality of chronic disease so that there has been an explosive growth in the world's population, particularly of individuals of advanced age – 75 years and upwards. Developments in public health measures and industrial hygiene have altered the environment to permit people to live in a more healthful condition, as well as permitting more people to live longer.

All of these aforementioned advances and many, many more have created practical ethical problems not only for health care professionals but indeed for all living human beings. Such questions – When does human life begin? What constitutes a human person? What forms of manipulation can human beings use on one another? – immediately come to mind in considering some of the advances above. There is no consensus among ethicists or persons who have given much thought to these and other fundamental questions raised by the new medical science and praxis because the ethicists and other individuals who have thought long and hard about these problems come from diverse backgrounds and with different culturally, psychologically, and religiously conditioned ethical stances.

Before commenting on the excellent contributions by Fathers Fuchs and Demmer and the responses by Professors Cahill and Hellwig, respectively, a few words are in order concerning the history of the Roman Catholic Church and its interest in what has come to be known as medical ethics.[1] Certainly by the middle of this century when the medical world was exploding with new scientific discoveries and technological advances, there was already a separate, solidly established, and well-developed discipline of medical ethics in the Roman Catholic tradition. This was demonstrated by the number of books on the subject published in English by, among others, Connell [9], Finney [11], Healy [15], Kelly [18], McFadden [21], and O'Donnell [23]; by Bonnar [6] and Niedermyer [22] in German; by the numbers of dissertations on medical issues from Catholic seminaries and pontifical universities; by Roman Catholic journals on medical ethics (e.g., in Germany *Artz und Christ*; in France *Cahiers, Laennec*; in Belgium *Sen Luc Medicale*; in England *The Catholic Medical Quarterly*; and in the United States *The Linacre Quarterly*).

One could ask why the Roman Catholic Church was concerned with this discipline when it did not exist for all practical purposes in any other religious tradition except the Jewish. There is a tradition of care for the sick in the Judeo-Christian tradition which was highlighted by the biblical history of Jesus' caring for the sick. The Church has always reflected on different aspects of sickness and the meaning of the reality of sickness for Christians. God, the author of life, has healing power. Sickness may be a sign of human weakness, futility, and frailty. Sickness and suffering of individuals have always been seen in the context of the suffering, death, and resurrection of Jesus, the created man, the anointed one of God and at the same time God, the uncreated Logos of the Father with His redemptive-resurrection destiny. The tradition of the Church has always included the spiritual care of the sick. The anointing of the sick as part of the sacramental mystery was described in the Epistle of James and has been practiced from Apostolic times.

A fundamental characteristic of Roman Catholic thought that carried over into Church's concern for medical ethics is the emphasis on mediation. God ordinarily works through secondary causes – mediately – thus, God works in this world through His chosen people, Israel, through the Prophets of the Old Testament, culminating in the revelation of Himself in His son, Jesus, the anointed one of God, the Christ, who established the Kingdom of believers in which God is present and comes to us through this visible community of believers, the Church. Hospices, hospitals, and other caring institutions were established by the Church.

Roman Catholic self-understanding stressed not only faith but good works. Neither one nor the other was more important. Both were stressed. Good works were a state of life, thus the tradition of encouraging activities of doctors, nurses, and other health care workers. In addition, Catholic moral theology has always been closely connected with penitential aspects of Catholic life. There has been great emphasis on the tradition of what is sinful and what is not, and a confession of sin and reconciliation of the penitent with one's fellow Christians. The Church interested itself in the visible community of the Kingdom on earth. This ecclesiological institution itself needs law for a universal, viable society.

The Church in its concern for the sacramental life of its members, such as in the matters of marriage and procreation, recognizes that the problems of impotence and sterility require study and medical knowledge. It also requires laws for the community concerning the validity of the sacrament. The Church's insistence on the sacrament of Baptism by water or desire led to study of the unborn child who might die in the womb and be in need of Baptism. This brought to light the field of embryology. The Church understood that the purpose of medicine was to cure and to preserve the life of the individual. The good of the individual was the purpose of medicine. Thus, good medicine and good morality were the same. Before 1950, the approach of medicine was to cure the individual and overcome sickness and suffering. With the technical advances since 1950, medicine has been able to interfere to yield a better individual. Thus, the Church began to be concerned with human reproduction but valued sexual function as existing for the good of the individual and the good of the species. There was a two-fold finality of the sexual organs. Conflicts arose over the individual aspect and the species aspect of sexuality. Likewise, the injunction "Do no harm" brought conflict concerning organ transplantation. Could one, in transplanting organs, harm one individual for the good of another? Therapy is good for the individual person. Medical experimentation may not be good for the individual person in the traditional sense of curing and overcoming an individual's sickness and suffering, but it may be undertaken primarily for the good of the species.

Throughout its history, the Church has been concerned with medical problems. In the Patristic period up to the year A.D. 600, the Fathers of the Church gave advice of a pastoral practical nature in their writings and sermons. The Didache [10] condemned abortion. Augustine [4] and Jerome [16] dealt with the infusion of the soul into the body – the question of when life begins. Clement of Alexandria developed his procreative rules and the Epistle of Barnabas [5] dealt with contraception

– concepts of sexuality derived from the pagan philosophers' concept of "Apaeithia" and found by them not to be specifically forbidden in the Gospel narratives. From the 8th Century onwards with the Church's concern with penance and reconciliation of the sinner with the community, penitential books were developed, dealing with the list of penances given for various sins dependent on their gravity. This activity culminated in the development of Canon Law, the codification of the laws of the community and the Church, to get people to live a better Christian life. In A.D. 1140, the Decree of Gratian [12] appeared which begins the Canon Law. Popes began to publish their own collections of Canon Law. Gregory the Ninth [13] in A.D. 1234, spoke of the need for medical knowledge to prove impotency as an impediment to marriage. In A.D. 1331, Pope John XXII [17] gave a collection of laws to the judges of the ecclesiastical law court, "The Rota", which he formed. He spoke of knowledge of medical skills needed for decrees of "The Rota". In the 13th Century, the brilliant Scholastic philosopher and theologian, the Dominican Friar Thomas Aquinas, produced his *Summa Theologicae* [3], a systematic work on theology that included moral theology.

Thomistic thought was interested in ends. Aquinas took the work of the pagan philosopher and natural scientist, Aristotle, with his concern for "telos" and explained this philosophy in terms of the Church community's understanding of the Gospel message of Jesus Christ. Thomas theorized a plan in the mind of God for His creatures – the divine or Eternal Law. As part of the Eternal Law there was Natural Law or "right reason", that is, reason directing us to our ultimate end in accord with our nature. This divine plan or Eternal Law is not written. Truth may be found, Thomas asserted, on the basis of human reasoning reflecting on human nature in light of faith. His work was a systematic treatment of all theology. The second part of the *Summa* dealt specifically with moral theology.

From Thomas' time down to the 18th Century, moral theology gradually lost its connection with philosophy and systematic theology and became a separate discipline. Where Thomas had emphasized "right reason" as an intrinsic property natural to the person, based on his creation by God and redemption by Jesus Christ, later moral theologians laid heavy stress on morality as conforming to the law "out there", or external to the individual. The morality of individual acts was determined in conformity to external norms or laws. Canon law and casuistry were synthesized.

In A.D. 1459, another Dominican Friar, Antoninus of Florence[2], produced a summary of the rules for moral life. Antoninus, Archbishop of Florence, involved in the day-to-day life of the Church, summarized the rules by which individuals were to live their lives. The third volume of his *Summa* dealt with medical ethics. He dealt with the competency of physicians, their diligence and practice, the practical considerations in dealing with dying patients, fees for medical service, and ways to deal with incurable patients.

In A.D. 1621, Paulo Zachia[24] produced the first Christian book devoted purely to medical ethics, *Quaestiones Medico-Legales*. He brought together for the first time questions of both legal and pastoral importance in medical practice.

In the 17th and 18th centuries moral theology considered what an individual should do in cases of doubt. Can the penitent in doubt follow a course when his confessor does not agree with that course? This was a struggle over probabilism. Is it permitted to follow a probable opinion in freedom from the law when there exists a more probable opinion in favor of the law? The controversy led to two extremes of moral behavior, rigorism (Jansenism) and laxism. The controversy lasted over two centuries and was finally settled at the end of the 18th Century, aided greatly by the writings of the founder of the Redemptorist Order of priests, Alphonsus Liguori[20], who adopted a moderate stance in the matter. The termination of the controversy ended extremes of laxism and rigorism and heightened the emphasis on law as the primary ethical determinant. Morality was identical to obedience to the law! It emphasized extrinsicism, for the solutions were prefabricated from outside the individual, personal responsibility was minimized, and great emphasis was placed on consideration of acts in and of themselves, particularly biological actions.

In the 19th and 20th centuries pastoral medicine, or medical ethics, developed in the Church. Carl Capellmann [8], in 1877, wrote *Pastoral Medicine* to instruct the pastor in medical knowledge to be used in carrying out his ministry and acting as a confessor, and to instruct the doctor by giving him principles of moral theology to enable him to practice his profession. Joseph Antonelli wrote *Medicina Pastoralis* in the late 19th Century[1]. Patrick Finney wrote *Moral Problems in Hospital Practice* in 1922[11]. In the 1950's a flood of books appeared. These were in essence manuals of medical morality which defined acts that were sinful in medical practice[19]. These manuals were aimed at assisting health care professionals in their practice. Pope Pius XII during the 1950's in allocutions and letters dealt frequently with the morality of medical praxis.

After 1950, moral theology became more life-centered. This anthropocentric approach was emphasized by Karl Rahner, the renowned modern dogmatic theologian. Bernard Häring at the Alphonsium University in Rome wrote his *Law of Christ*, an influential moral theological work, in 1954[14]. This publication emphasized that moral theology was life-centered. The essays by Fathers Fuchs and Demmer and the comments on them by Professors Lisa Cahill and Monica Hellwig, respectively, reflect this latter development in moral theology and serve to explicate the theological background underlying medical ethics in the Catholic tradition.

Where does the Catholic find ethical knowledge and wisdom? Where does the Catholic physician find answers to medical ethical questions? Father Fuchs considers the answers to these questions in his paper "Catholic' Medical Moral Theology?". He brings to our attention the Thomistic view of Natural Law, but leavened by a 20th century anthropocentric approach – a life-centered approach reflecting the work of the Jesuit theologian Karl Rahner. St. Thomas said that on the basis of human reason reflecting on human nature, wisdom and truth could be found. Fuchs says that "the human person understands himself as a moral being and necessarily understands medical conduct and action as moral conduct and action. Since medical problems are human problems, they are moral problems." Thus, Fuchs holds that medical ethics is based on philosophical ethics. Professor Cahill concurs, stressing that a key characteristic of the Catholic Natural Law tradition is to deny that any religious commitment or any theological anthropology is a necessary prerequisite for general moral insight. She does point out, however, that Christian authors operate within the general context of Christian commitment and theology so that any simple separation of faith foundations from substantive natural law conclusions is bound to be artificial.

She further observes that the representatives of Christian Natural Law have been insufficiently conscious of their own historical roots. Specific moral proposals, derived within a particular and necessarily circumscribed cultural, philosophical, and religious tradition, have been put forward authoritatively in Catholic teaching as the rational and immutable requirements of human nature as such. She holds that analyses in Catholic Natural Law require examining its philosophy more directly in relation to the religious and thological aspects of the tradition, and considering the degree to which these aspects have helped shape conclusions in Natural Law.

Fuchs, in supporting the concept that medical ethics is based on a philosophical ethics, argues that medical ethics cannot appeal to faith as its starting point and ultimate basis. He insists that human self-understanding precedes every faith and every believing reflection. He insists that medical problems do not have a general moral solution because God, in whom we believe, has ordained things in particular fashions. Instead, questions about which medical conduct and actions are correct must be answered on the basis of the self-understanding of the human person. Ethics based on faith or theology alone cannot be understood humanly and thus would not be understood as morally obligatory. For this reason medical ethics is an autonomous ethics not imposed on the human person from outside himself. Thus, the Catholic Christian physician can have dialogue or discourse on medical ethical problems with physicians of other Christian denominations and, indeed, with physicians of non-Christian faiths, because Moral Natural Law is the basis of medical ethics.

The human person must explore and discover, Father Fuchs insists, He indicates that the tragedy of the person is that he can go astray and arrive at false solutions, and the tragedy of humanity is that different people in search of truth can arrive at different solutions to the problems presenting themselves. He emphasizes that medical ethics is not found in God, nor in scripture, nor in Jesus Christ, nor in theology, but in reason directing us as human beings to our ultimate end in accord with our nature.

The God of faith, however, is not irrelevant, Fuchs contends, in the attempt to find solutions. The God who created and redeemed each one of us wills our search to be a search that provides solutions that truly correspond to our humanity, created freely as a loving gift of a totally sufficient and self-fulfilled God. Fuchs argues that medical ethics is theological if it proceeds from faith, which he holds is not an assertion of the truth of certain propositions but an act deep in the person of entrusting himself to God who reveals and imparts Himself to us. God seeks our salvation and has given us redemption, and this concern for us is that we be open to others, be open to all that is good, and not be closed in on ourselves. Jesus' way in the world was doing good, helping others, and healing. Thus, the human attempt to develop oneself in relation to medical ethics can be an attempt that derives from faith. Living faith signifies an interior disposition to seek truth. Such a disposition founded on the basis of faith has primary significance from the seeking and discovering of medical/moral norms. Thus, Fuchs says, the medical ethics of the Catholic physician who holds onto his faith in his moral reflection can be called a Catholic medical ethics. Faith and theological anthropology cannot replace human philosophical reflection in the seeking and discovering of truth. It is also true, Fuchs asserts, that, in seeking a Catholic medical ethics on the basis of faith, Catholic physicians may not all arrive at the same conclusion, for medical ethics are not revealed; they are not objects of faith but must be discovered "in the light of the Gospel". An individual who understands human existence from this faith standpoint may more easily discover right action than one who does not have such a base, Fuchs insists. As Cahill summarizes Father Fuchs ideas, biblical religion and faith act primarily in the role of motivators, and natural reason acts as a practical guide.

Fuchs goes on to deal with the Catholic tradition in medical ethics. He points out that while revelation and faith are given, theology is a basis for reflection on faith. Since theology is a human reflection, it is conditioned by human elements and this reflection can arrive at different theologies. But there are certain fundamental truths of faith relevant to ethics. They are: (1) God has created human beings. (2) He has taken all humans into His life. (3) All humans share in His love. (4) The dignity of all humans lies in His love. (5) All must acknowledge this dignity in all people. (6) All must be open to the other person.

When such truths are to be made concrete, human reflection begins. In the past it was thought that such reflections were made in the light of the Gospel and under the guidance of the Holy Spirit and thus were free from error. Today we have a different view of moral solutions within the Christian tradition. Catholics have given different answers to different questions at different periods in history. Father Fuchs sums this up by noting that neither faith nor dogmatic theology can provide concrete solutions to concrete moral problems.

Father Fuchs points out that early Christians adhered to evaluations and norms given by the pagan philosophers if they found no incompatibility between the moral norms given by these philosophers and the fundamental attitude of the members of the Apostolic community as expressed in Scripture. For example, the Stoics taught that the given nature of the human person should be understood as the norm for moral conduct. Within the Christian perspective it would then follow that nature indicates the divine will, since human nature is the work of the Creator. Thus, biological action becomes very important for moral norms. As a result, as Dr. Cahill points out, members of the Apostolic community expressed the meaning of Christianity within views of the world that differ quite markedly from own, which has a less teleological understanding of biological questions. For instance, the earliest communities expected Jesus to return as Lord within their own lifetimes. This expectation colored their ideas about the behavior expected for both personal morality and social responsibility. Today this New Testament eschatology does not have the same significance for Christian ethics. Natural Law moral theologians have tried to demonstrate a basic compatibility of Thomistic philosophical ethics with the Gospels, and Cahill reminds us that scripture scholars pose the question whether Christianity can be understood at all if ethics is taken as an autonomous object of interest rather than understood within the context of Discipleship.

Fuchs does not make this point. He points out that if neither faith nor dogmatic theology can provide concrete solutions to concrete moral problems and can do no more than preserve the light of the most general principles, then the concretization of such principles in the concrete circumstances of life must take place in another way. That way is by means of the evaluation and moral interpretation by the Chritian as a human person of the given human realities. This way is available even to the non-Christian and has been called "The Moral Natural Law". The traditional interpretation of Moral Natural Law is conditioned by history and experience, since it is always a human attempt and thus subject to error. Thus, Catholic medical moral theology must not accept traditional teaching blindly at all points.

Father Fuchs deals with the Church's Magisterium, the authentic teaching body composed of Pope and Bishops. He points out that within the Church there is an official and valid presentation and explanation of tradition. Tradition is understood in relation to the doctrine of the faith of the Church. Teaching and ethics are seen as standing in ultimate relationship to the doctrine of the faith of the Church. As guardian of the faith, the Church intervenes in official ways in moral theology. The doctrine of faith does not contain answers to concrete moral questions. The Church itself generates them. This does not come from the treasure of the Church's faith. Rather, Fuchs asserts that there is no special knowledge in certain areas such as sexuality, embryology, genetic composition, and the like. But knowledge about these matters is presupposed in moral affirmations about human behavior. Father Fuchs points out that the formulations of both the first and second Vatican Councils indicate that questions of natural law are not bound to the Magisterium as are questions dealing with Christian revelation. Biologists and theologians feel bound for instruction and practices by these statements. But they are not infallible statements, though infallible statements include statements not found in the treasury of the Church's faith, but in statements derived from human reflection. The Church's Magisterium insists that one must adhere to such teaching, though dissent is not absolutely excluded.

Cahill points out that the Magisterium of the Church understands itself as having authority under the guidance of the Holy Spirit to interpret the moral implications of philosophy, scripture, and contemporary experience, and it does not see itself as equally subject to the critique of any of these other sources. She maintains that it is difficult to see exactly how the Catholic Church as a religious and theological community is to function authoritatively within the philosophical definition of moral obligation. When authoritative moral proposals are made, Cahill says it must be asked whether they can stand on philosophical warrants, and if not whether the theological reasons backing them have been explicated adequately.

She also points out that a new perspective on Natural Law has been gained since the second Vatican Council; it is understood within an historical religious community under the auspices of a formal magisterium. This new perspective includes the insight that knowledge of natural law is mediated through experience and that the evaluation of any individual act must take account of the social context of that act. This was addressed by Pope John XXIII when he called Catholic theology to encounter the "Signs of the Times". Cahill points out that Pope John Paul II has made persistent appeals to concrete experience in some of his remarks on sexual ethics, especially in the context of married love. Cahill also points out that the turn to concrete experience raises questions regarding the criteria for valid or normative experience. In this context she says that special attention should be given to persons who participate in the experience under evaluation, and that the witness of these persons should have some impact on the resulting normative conclusions. Cahill properly points out that Natural Law tradition with which we are most familiar has developed in the prosperous, post-Enlightment West; that we have to take into account the experiences of persons and communities cross-culturally; and that we must attend to the effect these considerations will have or should have on the way we institutionalize moral authority in our theological ethical tradition.

Klaus Demmer grounds his arguments in moral theology on the philosophically based 20th-century Thomistic-anthropological concept of Natural Law. He remarks that there is no recourse to the normative will of God, and that this approach was never seriously attempted in Catholic moral theology despite a contrary impression as a consequence of the paradigm of revelation theology. Demmer asks the central question: In what way and to what extent can epistemological discussion be incorporated into theological arguments? He points out that the sciences are in motion. They are an open system in continuous movement with a progressive accumulation of material so that a permanent critical correction of basic assumptions is necessary. Insofar as developments are not progressive, but discontinuous, the notion of a shift of paradigms is introduced. The problem of bridging the gap between empirical verification and theoretical understanding and interpretation then becomes particularly challenging. Such paradigm shifts take place when an old paradigm becomes sterile and a new one is judged more promising. In theology, a shift in paradigms provides an opportunity for the rediscovery of one's own hidden and perhaps forgotten tradition.

The criteria for the theological arguments are found in the nature of moral truth, the demands of life commitment and the fact that moral truth is projected by autonomous reason. Demmer holds that right reason can disclose the whole and full meaning of moral values. The theology of revelation indicates that Christian revelation does not provide man with operative or moral norms. It provides the believer with self-awareness and self-understanding on the level of operative moral norms. Anthropological norms indicate the uniqueness of man, man's radical equality, and a new meaning of death. One cannot deduce moral norms directly from revelation or faith. There is, however, a process of self-understanding, a process of profound hermeneutics for the Christian believer that leads to operative moral norms. Father Demmer discusses the historicity of this process of hermeneutics and how man discovers normal norms, revises them, and discovers them anew. This is a creative process and is open to new experiences and even to shifts in paradigms.

The second point that Father Demmer makes is that we must recognize paradigm shifts on the level of arguments in Natural Law. One must keep in mind the dialectic on the level of Natural Law. The first element of this is that the human mind is in a position of continuous learning about nature. Man enriches his self-understanding. The second element is that each person must try to evaluate this permanent process of learning. Natural Law is in permanent movement. Moral reasoning tries to understand and interpret natural data in the light of basic anthropological assumptions.

Classical moral theory is infused with new basic scientific knowledge. Demmer points out, for instance, that there has occurred a reinterpretation of the principle of totality. The teaching of Pius XII began as a reductionist teaching. Later, a broader principle of totality led to support for organ transplantation. This involved a shift in paradigms. Likewise, there is a shift involved in considering in-vitro fertilization in a heterologous system. The question is whether this constitutes a true paradigm shift. There is tension between individual good and common good. Social structures of family and marriage are involved. There might come a new understanding of personal good and the structure of society. He points out also that our understanding of person and substance in light of the classical definition of Boethius is changing. What do substance and person mean in the light of modern physics and biology?

Dr. Hellwig in her reply brings up an unrecognized question: whether there truly is a role for moral theologians in medical ethics and bioethics as distinct from ethics pure and simple. She asks what defines these roles and how they get their shape and form. She points out the scope and limitations of theological argumentation in the context of revelation: revelation is found in biblical texts and revelation is in magisterial documents. Hellwig indicates that two levels of hermeneutics are addressed in Father Demmer's paper: the more superficial one lets the text speak, but hermeneutics places the whole in the context of our understanding of revelation and magisterial texts. These latter have to be put, then, into their historical and cultural contexts. She asks, "What do we do with the present magisterial texts that claim no paradigm shift has occurred and that there never will be a paradigm shift?" She points out that Father Demmer localizes the problem of hermeneutics so that anthropology becomes the contact point for the sources of inspiration and argumentation on specific issues.

Hellwig insists there is a second aspect that is treated in Demmer's paper: the handling of the Natural Law arguments in the contemporary Catholic tradition. Natural Law to outsiders is preconceived as an extremely narrow confessional project. Others maintain that Roman Catholics are not discussing this in the open. We find ourselves affected by two sets of paradigm shifts. One is within the faith tradition and the other is within the secular world, brought about by technology and cultural cross currents. She asks what criteria we as Roman Catholics have to scrutinize these paradigm shifts.

There are two major issues concerning these shifts, she asserts. One is the question of the justification of any paradigm shift and the other concerns methodologies to bring shifts within disparate contexts into one focus. The goals are dialectic, and involve an exchange between theory and praxis both within the world and within religious contexts. There is a continuous demand to sort out and reshape our understanding of human life regarding creatures, caring, etc. The theological aspect of this demand is the need of criteria for scrutiny. The criteria offered are certain themes in revelation concerning creation and redemption. Self-understanding precedes the faith commitment. Hellwig points out that Fuchs has also insisted on this and that our self-understanding is constantly being formed within a dialectical understanding of reality. All factors play a role in shaping our understanding, and faith does not come last! In this she disagrees with Father Fuchs. The heritage of sin in the world is no easy guarantee that we are seeing things as they really are. It is redemptive to think clearly.

Hellwig insists that we must listen to everyone and reflect on what we see and why we see it. We should be careful not to construct Natural Law arguments behind closed doors. We are trying to catch up through a mixed reflection on patristic and revelatory reflections. The Catholic Church has a strong magisterial teaching role which makes demands in conflict with studying Natural Law with non-Catholics. Hellwig compares this current magisterial concept of nature and Natural Law to the theological reflection presented by Father Demmer, one which incorporates an anthropological approach to understanding reality and human life. She assists us in clarifying the tensions existing between these two approaches.

We thus come to see that the Roman Catholic Church (from Apostolic times to the present) has always had an interest in those matters of the human condition which emerge from illness and physical suffering. The rightness or wrongness of actions on the part of patients and health care workers in dealing with these matters (medical ethics) has always been judged by the Church on the basis of human reason reflecting upon human nature in the light of the faith expressed by Jesus, the Christ, true man and true God, in the Gospel message of love. Right reason was defined by Saint Thomas as part of the Natural Law. Throughout the ages this judgement, though always based upon Natural Law, has been conditioned by cultural, philosophical, and religious influences. In contemporary medical ethics, the Church's stance is still essentially Thomistic, but it is influenced now by an anthropomorphic life-centered approach emphasized by Father Karl Rahner, perhaps the most influential Jesuit dogmatic theologian of this century, and endorsed by the Fathers of the second Vatican Council.

Fathers Fuchs and Demmer tend to emphasize the consistent theological background for clarifying current issues in medical ethics in the context of our 20th Century, life-centered approach. Their explanations are further clarified by the comments of Professors Cahill and Hellwig. The four preceding essays, then, serve as very concise yet helpful analyses of the theological basis for medical ethics within the Roman Catholic tradition.

School of Medicine
Georgetown University
Washington, D.C., U.S.A.

NOTE

[1] In this brief historical outline, I have relied greatly on numerous discussions with Professor Charles Curran in 1975 at the Catholic University of America.

BIBLIOGRAPHY

1. Antonelli, J.: 1932, *Medicina Pastoralis in usum confessariorum professorum theologiae moralis et curiarum ecclesiasticarum*, F. Pustet, Romae.
2. Antoninus of Florence: 1582, *Summae sacrae theologiae, iuris pontificii et caesarei*, Apud Juntas, Venetiis.

3. Aquinas, St. Thomas: 1947, *Summa Theologicae*, trans. the Fathers of the English Dominican Province, Benzinger Brothers, Inc., New York.

4. Augustine: 1950, *City of God*, trans. Marcus Dods, Random House, Inc., New York.

5. Barnabas: 1940, *Doctrina duodecim apostolorum: Barnabae epistula*, ed. T. Klauser, Florilegium Patristicum, Bonn.

6. Bonnar, A.: 1939, *The Catholic Doctor*, P. J. Kennedy and Sons, New York.

7. Clement of Alexandria: 1897, Paedagogus 2.10.96.1 in *Die Griechischen Christlichen Schriftsteller der ersten drei Jahrhunderte*, Leipzig.

8. Capellmann, C.: 1901, *Medicina Pastoralis*, Aquisgraum.

9. Connell, F. J., C.S.S.R.: 1951, *Morals in Politics and Professions*, The Newman Press, Westminster, Maryland.

10. Didache: 1947, *The Apostolic Fathers, The Fathers of the Church*, trans. F. X. Glinn, Vol. 1, Cema Publishing Co., Inc., New York.

11. Finney, P. A., C.M.: 1922, *Moral Problems in Hospital Practice; a Practical Handbook*, B. Herder Book Co., St. Louis, Missouri.

12. Gratianus: 1554, *Decretum Gratiani, universi iuris canonici pontificas constitutiones et canonicas brevi compendio completus*, I. Pidaeius, Lugduni.

13. Gregory IX: 1605, *Decretales de Gregorii Papae IX*, Juntas, Venetiis.

14. Häring, B., C. S. S. R.: 1961, *The Law of Christ*, trans. Edwin Kaiser, Newman Press, Westminster, Maryland.

15. Healy, E. F., S. J.: 1956, *Medical Ethics*, Loyola University Press, Chicago.

16. Jerome: 1942, *The Letters of St. Jerome*, ed. James Duff, Browne and Nolan, Ltd., Dublin.

17. John XXII: 1983, *Extra-agentes Joannis XXII*, ed. Jacquelna Tarant, Biblioteca Apostolica Vaticana, Citta del Vatican.

18. Kelly, G., S. J.: 1958, *Medico-Moral Problems*, The Catholic Hospital Association, St. Louis.

19. Kenny, J. P., O. P.: 1952, *Principles of Medical Ethics*, Newman Press, Westminster, Maryland.

20. Liguori, A. M., 1840: *Compendium Theologiae Moralis Sancti Alphonsi Mariae de Liguorio, sive, Medulla Theologiae moralis Hermani Busenbaum*, Iriae.

21. McFadden, C. J., O. S. A.: 1976, *The Dignity of Life: Moral Values in a Changing Society*, Our Sunday Visitor, Inc., Huntington, Indiana.

22. Niedermeyer, A.: 1956, *Compendium der Pastoral – Hygiene*, Herder, Vienna.

23. O'Donnell, T. J.: 1956, *Morals in Medicine*, The Newman Press, Westminster, Maryland.

24. Zacchia, P.: 1621, *Quaestiones Medico-Legales*, Amstelaedami, Romae.

PART IV

PLURALISM WITHIN THE CHURCH

RICHARD A. McCORMICK

PLURALISM WITHIN THE CHURCH

Let me begin with a citation from *Civiltà cattolica*:

Catholic principles do not change either because of the passage of time, or because of different geographical contexts, or because of new discoveries, or for reasons of utility. They always remain the same, those that Christ proclaimed, that popes and councils defined, that the saints held and that the doctors defended. One has to take these as they are or leave them. Whoever accepts them in their fullness and strictness is Catholic; whoever wavers, drifts, adapts to the times or compromises can call himself whatever he likes, but before God and the Church he is a rebel and a traitor ([5], pp. 145–159).

These words were written in 1899 as an editorial commentary on the condemnation of Americanism. They could well have appeared in last week's *Wanderer* or *National Catholic Register*, for they are a symbol of the Catholic integratist mentality. For such a mentality the very title of my paper does not represent a question; it represents an abominable error and even a heresy.

I mention this at the very outset for two reasons. First, it is not the way this discussion ought to be conducted. Second, it is unfortunately the way it is frequently conducted. Daniel L. Donavon summarizes many of the discussions during the Modernist controversy as follows:

The task of understanding was made more difficult by the use of stereotypes and generalizations. Recourse was constantly being had to "isms" of every kind. Blondel's *L'action*, for example, was condemned as Kantianism, psychologism and subjectivism. References to life and experience were rejected as fideism, false mysticism and pragmatism. Laberthonnière repudiated scholasticism under whatever form as intellectualism, and Tyrrell called the system that challenged him Vaticanism, Jesuitism and Medievalism. The atmosphere, in short, was not conducive to either understanding or discussion. The tendency to polarization was an important factor in all that happened ([5], p. 155).

With a change of a few scarlet words that paragraph could be written today about moral theology. The likely candidates for inclusion are: subjectivism, absolutism, situationism, dualism, dissentism, utilitarianism, biologism, consequentialism, deontologism, rationalism, etc. When theological issues get trapped in such language, they are usually suppressed and not faced squarely. When that happens, they return to haunt, harass, and hurt us at a later date. If the Modernist crisis teaches

147

Edmund D. Pellegrino et al. (eds.), Catholic Perspectives on Medical Morals, pp. 147–167.
© *1989 Kluwer Academic Publishers. Printed in the Netherlands.*

us one thing, it is that. If we think we "solve" genuine theological issues with mere formal authority, the problem remains unsolved. For this reason, M. Petre has noted that "had it been feasible for the different sections of modernism to unite in the insistence on one point, which should be vital to all, that point would have been the character and limits of ecclesiastical authority" ([14], p. 141).

I want, therefore, to discuss pluralism in the Church without surrounding the idea unduly with other "isms". I will proceed through five points: (1) Areas where there is no disagreement; (2) Areas of dispute; (3) How the issues get confused; (4) Areas of pluralism in medical ethics; (5) Some personal procedural suggestions on pluralism.

I. AREAS WHERE THERE IS NO DISAGREEMENT

There are two areas here. First, all Catholic theologians would agree that there can be no pluralism at the level of universal principles and formal moral norms. By "universal principles" I mean generally stated moral norms that impose achievement of a value or proscribe a disvalue. An example would be: there is always a presumption against taking human life. Under the term "formal moral norms", we may include two types of statements. First, there are normative statements such as "our conduct must always be just". Second, there are normative statements that build around words that include their own value judgments. For example, "we must never commit murder" (murder – *unjustified* killing). Equivalent to this type of statement is one that exhaustively states the circumstances (e.g., it is morally wrong to kill a human being *merely to give pleasure to a third party*). People who advocate pluralism in these types of statements either do not understand them or have placed themselves beyond civil moral discourse.

The second area where there is agreement would be the area of acceptable pluralism. All Catholic theologians would accept pluralism in the following matters.

(a) The Application of Universal Principles to Contingent Facts

The American bishops in their pastoral *The Challenge of Peace* explicitly acknowledge the acceptability of pluralism here. Examples would be: the moral legitimacy of capital punishment, the morality of any (or first) use of nuclear weapons, whether a particular sterilization is direct or indirect.

(b) Emphases in Issue Areas

Human beings are limited. Very few can be expert in a great many areas. That means that all of us must specialize, pick and choose our areas of concern according to our talent, competence, and interests. Such specialization necessarily results in a kind of pluralism of focus and concern. No one questions the appropriateness of this pluralism within the Catholic idea.

(c) New Problems

In contemporary medicine, we experience almost daily the casting up of new and complicated questions. What is the most equitable approach to broaden access to health care? When it is morally justifiable to begin genetic therapy for single-gene defects? How should we analyze artificial nutrition-hydration for the permanently vegetative? When is joint-venturing likely to compromise the Catholic character of a health facility? Is the artificial heart, all things considered, a justifiable therapeutic modality? The list is endless. On many such questions – especially when they trace to new technology – there is little experience or reflection to draw on. One expects pluralism in such areas.

(d) Disputed Questions

This is close to the third group but it does not perfectly overlap. I have in mind problems that have proved especially resistant to unanimous resolution within a believing community. A clear set of examples would touch on the best legal policy on divisive questions such as abortion, surrogate carriers, genetic screening, homosexual rights. Other examples would concern adolescents, their sex education and access to certain forms of health care. Such questions are often quite complicated and demand empirical knowledge from a variety of sources. We expect a genuine pluralism on questions like this and generally abide it rather well.

II. PLURALISM: AREAS OF DISPUTE

By "areas of dispute" I mean areas where some believe pluralism is acceptable, others that it is not. I can identify three general areas.

(a) Concrete Moral Norms

The obvious examples here are some of the following. Every direct abortion is morally wrong. Artificial insemination by husband is always morally wrong. Masturbation, regardless of circumstances, is intrinsically evil. Every conjugal act must remain open to the possibility of procreation. Direct sterilization is always morally wrong.

The five examples of norms that I have given have all been proposed in authoritative fashion by the papal magisterium. Further examples could be cited. Furthermore, all of them have been contested in the moral theological literature of the past twenty years. Some find such contestation intolerable: it undermines authority, destroys unity, creates a theological para – magisterium, confuses the faithful, and eviscerates morality into an adjustment to the comfortable.

Others take an entirely different point of view. Far from undermining authority, dissent on such issues, when well argued, is a high form of loyalty to the magisterium, by protecting it from the kind of self-preoccupation that subordinates truth to authority. As for unity, Catholic unity should not be staked on such details but on the true substantials of the faith. A para – magisterium is not created because it is of the very nature of theology to exercise a critical role. To negate this critical role is to conceive the magisterium in an utterly pyramidal, other-wordly, and magical way. If the faithful are confused by this, it is because they have been misled by past attitudes and practice into viewing the magisterium in simplistic and unrealistic ways. Correction of these attitudes, not denial of theology's critical role, is in place. This is the way the discussion has gone and I shall return to it below.

(b) Methodology of Moral Norms

Within the past twenty years, many Catholic moral theologians have adopted a form of teleology in their understanding of moral norms. It is impossible in a brief space to give an adequate summary of this development or an adequate account of the differences that individual theologians bring to their analyses. However, common to all the analyses is the insistence that causing certain disvalues (ontic, non – moral, pre – moral evils) in our conduct does not *ipso facto* make the action morally wrong. The action becomes morally wrong when, all things considered, there is no proportionate reason justifying it. Thus, just as not every killing is murder, not every falsehood a lie, so not every artificial intervention preventing (or promoting) conception is necessarily an unchaste act. Not every termination of a pregnancy is necessarily an abortion in the moral sense. This has been called "proportionalism" especially by those who resist and reject the *Denkform*.

The point to be made here, however, is that those who resist this teleological analysis of moral statements do not only reject it. They regard it as incompatible with what they call "received teaching," even with revealed morality. Clearly, something incompatible with revealed morality is not a matter of free discussion – pluralism – in the Church.

Those who have proposed such a teleological foundation for the interpretation of moral norms obviously take a different point of view. They believe that such an interpretation is not only compatible with the abiding substance of Catholic teaching, but actually better accomodates and explains it. For instance, they argue that only a form of teleology (what Bruno Schüller calls "restrictive interpretation") can adequately account for the formula "no *direct* killing of *innocent* human life." Depending on the subject matter, such a teleological structure for moral norms would lead to a rewording of certain past formulations, doctrinal development in some areas, dissent with regard to certain conclusions (e.g., the intrinsic moral evil of every contraceptive act), the abandonment of the rule of double effect as a critical analytic tool. All of these outcomes, however, are seen as quite compatible with, even demanded by an ethic that roots in the centrality of the person as the criterion of the morally right and wrong ([21], pp. 37–38). The New Testament, it is argued, did not present us with a fully developed ethical system. Therefore, the behavioral implications of our "being in Christ" could be systematized and detailed in a variety of ways – as indeed they have been over the centuries of Christian moral thought.

(c) Degree of Authority in Authentic Teaching

Once again, this is a matter of dispute. One school of thought – what I still regard as a minority view – contends that very detailed concrete norms (e.g., the prohibition of contraception) can be, and indeed are, proposed infallibly by the magisterium. This is the thesis of John C. Ford, S. J., and Germain Grisez ([7], pp. 258–312). They argue that what has been proposed unanimously by the magisterium (popes and bishops) over the centuries as a tenet to be held definitively is *infallibly* taught. It would be proposed, they argue, as a truth required to guard the deposit of faith as inviolable and to expound it with fidelity. At a meeting (April 7–12, 1986) of moral theologians held in Rome (described by Grisez as "a celebration of moral theology loyal to the Church's teaching authority"), Grisez delivered a paper proposing as definable the following proposition: "The intentional killing of an innocent human being is always a grave matter" ([19], p. 14). Clearly, then, authors like Grisez do not see the infallible proposal of concrete norms as an open question, one where pluralism might be appropriate.

Another school of thought – by far the vast majority of theologians– rejects this analysis on several grounds. First, it is one thing to teach something as involving a moral obligation. It is quite another to propose it as *to be held definitively* (sc., to give irrevocable consent). Second, there is little or no evidence that matters such as contraception are so necessarily connected with revelation that the magisterium could not safeguard and expound revelation if it could not teach such matters infallibly. Third, there is the question of the significance of unanimous episcopal teaching in some areas. In the deliberations of the Birth Control Commission, Cardinal Suenens noted:

We have heard arguments based on 'what the bishops all taught for decades.' Well, the bishops did defend the classical position. But, it was one imposed on them by authority. The bishops didn't study the pros and cons. They received directives, they bowed to them, and they tried to explain them to their congregations ([12], p. 170).

Finally, considerations such as these have led theologians like Karl Rahner to conclude that concrete moral norms are not the proper object of infallibility. As Rahner words it:

Apart from wholly universal moral norms of an abstract kind, and apart from a radical orientation of human life towards God as the outcome of a supernatural and grace-given self-commitment, there are hardly any particular or individual norms of Christian morality which could be proclaimed by the ordinary or extraordinary teaching authorities of the Church in such a way that they could be unequivocally and certainly declared to have the force of dogmas ([19], p. 14).

So far I have attempted to describe a factual situation as objectively and dispassionately as possible. My objective here has been accuracy, not precisely persuasion to this or that point of view. The same cannot be said for my third point.

III. CONFUSION OF THE ISSUES

Under this title I will list factors that I believe confuse the issue of pluralism in Catholic moral theology. By "confuse the issue" I mean "make it more difficult to identify the level of tolerable pluralism." Usually this "confusion of the issues" will take the form of an oversimplification in which pluralism will be presented as *being, involving, entailing, leading to* something that it need not be, involve, imply, or lead to. I have in mind, therefore, a kind of guilt by association. It is clear that here my own perspectives begin to play a more prominent role in the analysis.

(a) Fact and Value Words

When one intermingles these two indiscriminately as if there were no difference, then one whose analysis justifies a killing is seen as one who justifies *murder* (– unjust killing). If a *Denkform* does that, then clearly it represents an unacceptable pluralism of approach to the moral life.

Let me take but a single example here. Paul Quay, S. J., asserts that several moralists "have been seeking to eliminate 'absolutely binding' moral norms." His examples: defrauding laborers, adultery, abortion "and the like" ([17], pp. 347–364). There is a confusion here between fact-description (*Tatsachenbegriff*) and value description (*Wertbegriff*). "Defrauding laborers," like adultery, murder, theft, is a value-description; indeed, a morally perjorative one. To state the contemporary discussion as if it were an attempt to justify what has already been labeled as morally wrong is to miss the point and engage in homiletics – and indeed the kind that condemns *ab initio* attempts to qualify past formulations. In other words, it blackens pluralism of approach by associating it – inaccurately – with moral horrors we would all disown. Such *ignorantia elenchi* (missing the point) only confuses the issue of legitimate methodological pluralism by misrepresenting it. This has happened so often in recent years, and even in high places, that the term "pluralism" itself has become suspect.

(b) Teleology vs Deontology

At times differences in moral theology in the Church have been formulated in terms of teleology and deontology. I do not believe that this has served well the cause of discovering an acceptable level of pluralism, and for many reasons. They are sprawling terms that are variously understood and as such allow people to talk past each other, making assumptions and attributing positions that are inaccurate or downright false. Thus, for instance, some Catholics who identify themselves as deontologists have accused others identified as teleologists of holding that "a good end justifies a morally evil means." No reflective theologian would or could hold such a thing; for it an action is said to be morally wrong, nothing will justify it. What has happened here is that certain actions that a Catholic deontologist sees as morally evil (e.g., contraception), a teleologist does not. The discussion should center around why certain actions are morally wrong or not, and whether and why disagreement on such judgments is acceptable or unacceptable pluralism. If it escalates beyond this, it tends to take on the character of denunciation, rather than analysis. Thus Schüller notes:

> It is time, then, for the deontologists and teleologists finally to give up the business of discrediting each other with moral verdicts. Whether a teleologist accuses a deontologist of 'worshipping the law,' or a deontologist slurs a teleologist by making him assert that a good end justifies every means, in both cases nothing has been produced that could contribute to an objective clarification of the controversy ([23], p. 168).

When nothing is produced, the place of pluralism is not clarified. In this sense, teleology and deontology are about as revealing as "liberal" and "conservative".

(c) The Pairs Right-Wrong, Good-Bad

This terminology is borrowed by some contemporary Catholic theologians from Anglo – American philosophy. Moral goodness refers to the person as such, to the person's being open to and decided for the self – giving love of God. It is the vertical dimension of our being. It is salvation. Another level is the horizontal. This refers to the proper disposition of the realities of this world, the realization in concrete behavior of what is promotive for human persons. We refer to this as the rightness or wrongness of human conduct.

The discussion of moral pluralism in the Catholic community is concerned with the rightness-wrongness of human actions. For instance, is it unacceptably divisive for Catholics to disagree about the moral character (rightness-wrongness) of *in vitro* fertilization? And why? Failure even to make the distinction between right-wrong, good-bad means that the moral life can be absorbed into discussions of goodness-badness (involving intention, inclination, goodwill, the virtues, etc.). This is the key error in the work of Servais Pinckaers, O. P. It is responsible for his mistaken assertion that agape is not functional in so-called "proportionalist" thought ([15], pp. 118–212). Agape is simply not the issue under discussion. It is responsible for his misleading assertion that faith and the gospel must have first place in Christian morality. Of course they must; but that is not the issue. The question is: What do the faith and the gospel concretely demand of followers of Christ, and not merely in terms of sentiments and desires? The answer to that is a question of rightness and wrongness. Pinckaers fails to see this. The result is that he confuses the issue of pluralism by raising it in terms that have nothing to do with it. One does not clarify the nature and rules of baseball by emphasizing certain dimensions of football.

(d) Substance and Formulation

In contemporary moral writing this distinction has functioned far less than I think it should. Following John XXIII, Vatican II stated:

Furthermore, while adhering to the methods and requirements proper to theology, theologians are invited to seek continually for more suitable ways of communicating doctrine to the men of their time. For the deposit of faith or revealed truths are one thing; the manner in which they are formulated without violence to their meaning and significance is another ([1], p. 268).

If there is a distinction between the deposit of faith and its formulation at a given time, this is *a fortiori* true of the behavioral implications of this deposit. For behavioral implications are even more dependent on the contingencies of time and place. To restate the Council's distinction in theological shorthand, we may distinguish between the substance of a moral teaching and its formulation.

If this distinction is not made, then certain kinds of disagreements among Catholics (pluralism) will be viewed as an attack on and corrosive of the substance of the Church's moral concerns and intolerably divisive. I believe that this happens frequently in the contemporary Church.

A few examples from bioethics can serve as illustrations. In 1986 the American Medical Association released its guidelines on treatment of the dying. The A. M. A. statement accepted the moral acceptability of withholding or withdrawing artificial nutrition-hydration from those in a persistent vegetative state under certain conditions. Archbishop Philip Hannan (New Orleans) stated in his diocesan paper that "the Church strongly condemns this position"[11].

Whether such a position is "contrary to the Church's teaching" depends entirely on what one conceives to be the Church's teaching in this area. Pius XII, following the theology of his time, formulated that teaching in terms of the limits on the duty to preserve life. Furthermore, such limits were stated in terms of ordinary and extraordinary means. Pius XII acknowledged the great relativity (to time, place, circumstances, etc.) involved in fleshing out these notions in practice. What is ordinary care (and obligatory) for one time, place, patient might not be in altered circumstances.

So far so good. But when we range over a number of decades to see how these concepts were variously applied and lived out, we are faced with the question: "What is the Church's teaching?" John Connery, S. J., has answered this question as follows: The Church's teaching forces us to judge these matters in terms of the *quality of treatment* (its burden or benefit) not the *quality of life* treated. Therefore, he concludes with Hannan, we may not withdraw artificial nutrition-hydration from the permanently comatose on the basis of their quality of life alone ([4]) pp. 26–33).

I disagree with that analysis and I would argue that the Church's substantial teaching does not impose the hard-and-fast distinction between quality of treatment and quality of life that Connery posits. I would argue that the Church's teaching (its *substantial* concern) is as follows: *life is a basic good but not an absolute one and therefore there are limits on what we must do to preserve it.* That is the substantial judgment we must uphold and carry with us into a variety of changing circumstances and technological possibilities. Anything beyond that general judgment is changeable formulation. Thus, how we should formulate these limits (whether dominantly in terms of person or treatment), what we should call them, etc., do not pertain to the substance of Catholic teaching. Whatever the case, failure to make the distinction between substance and formulation will confuse the issue of pluralism by locating unchangeable Catholic teaching at a level where it might actually be changeable.

I can illustrate this point again by referring to Joseph Cardinal Bernardin's Gannon Lecture at Fordham (Dec. 6, 1983)([2], pp. 491–494) and his Wade Lecture at St. Louis University (March 11, 1984)([3], p. 705) on a consistent ethic of life. At one point Cardinal Bernardin refers to "The Challenge of Peace" and says there is found "the traditional Catholic teaching that there should always be a *presumption* against taking human life. But in a limited world marked by the effects of sin there are some narrowly defined exceptions where life can be taken." So far, so good. At another point Cardinal Bernardin refers to the principle "no direct taking of innocent human life" and says that it is "at the heart of Catholic teaching on abortion" and also "the most stringent, binding and radical conclusion of the pastoral letter [The Challenge of Peace]: that directly intended attack on civilian centers is always wrong."

Here we have two different statements referred to as "principles". I want to suggest here that the presumption against taking human life is the principle (or substance) of Catholic teaching in this matter. The rule, on the other hand ("no direct taking of innocent human life"), is a kind of formulation-application of this substance. By that I mean that the rule has developed as a result of our wrestling with concrete cases of conflict where we attempt to provide for exceptions but at the same time to control them. Such concrete rules, being data-related, are somewhat more malleable than the substance and will not always share the same force or universality as the substance.

Questions like this might be raised in a number of areas of medical ethics. For instance, the conclusion "direct sterilization is intrinsically evil" is not a *necessary* envelope for the Church's substantial concerns in this area. Until we acknowledge this, the discussion of pluralism will remain confused.

(e) Situationism

I can be brief here. This sprawling term is often used to characterize the positions that provide for certain exceptions to otherwise binding normative statements. The term is used in a very pejorative way, presumably to underline the subjectivism and relativism its users reject in the positions they criticize. I have no doubt that there are certain metaethical positions that involve unacceptable components of relativism and subjectivism. But this should not be a pretext for use of a term to tar all approaches that provide for exceptions to normative statements sometimes regarded as absolutely binding. The reasons given for exceptions must be treated on their own merit. If they are not, certain forms of pluralism will be condemned before they have been examined. This only confuses the question.

(f) The Fonts of Morality

Occasionally the matter of pluralism is approached from the perspective of the fonts of morality. Thus some will assert that, according to Catholic teaching, certain acts are morally wrong *ex objecto*. Denial of this is seen as unacceptable departure from Catholic teaching, as illegitimate pluralism.

Let me use John Paul II as an example here. In his apostolic exhortation "Reconciliation and Penance" John Paul listed several influences that undermine the sense of sin in our our time. One such influence he identified as a "system of ethics". He stated: "This may take the form of an ethical system which relativizes the moral norm, denying its absolute and unconditional value, and as a consequence denying that there can be intrinsically illicit acts, independent of the circumstances in which they are performed by the subject"[16]. The Holy Father was, I believe, ill served by his theological advisors in framing the matter in this way.

Equivalently, the Pope is saying that certain actions can be morally wrong *ex objecto* independently of the circumstances. But, as Bruno Schüller, S. J., has noted, that is analytically obvious *if the object is characterized in advance as morally wrong* ([22], pp. 535–559). No theologian would or could contest the papal statement understood in that sense. But it is not the issue. The problem is: what objects should be characterized as morally wrong and on what criteria? Of course, hidden in this question is the further one: What is to count as pertaining to the object? That is often decided by an independent ethical judgment about what one thinks is morally right or wrong in certain areas.

Let the term "lie" serve as an example here. The Augustinian-Kantian approach holds that every falsehood is a lie. Others would hold that falsehood is morally wrong (a lie) only when it is denial of the truth to one who has a right to know. In the first case the object of the act is said to be falsehood (– lie) and it is seen as *ex objecto* morally wrong. In the second case the object is "falsehood to protect an important secret" and is seen as *ex objecto* morally right (*ex objecto* because the very end must be viewed as pertaining to the object).

These differing judgments do not trace to disagreements about the fonts of morality (e.g., about the sentence "an act morally wrong *ex objecto* can never under any circumstances be morally right"), but to different criteria and judgments about the use of human speech, and therefore about what ought to count as pertaining to the object. In this sense one could fully agree with the Pope that there are "intrinsically illicit acts independent of the circumstances" and yet deny that this applies to the matters apparently of most concern to him (sterilization, contraception, masturbation).

In summary, then, if moral theological pluralism is discussed in terms of the fonts of morality, it is likely to be confused, not clarified.

(g) Subjectivism

There are those who shortcircuit the discussion of pluralism – thereby only confusing it – by describing departures from official formulations and the analyses that buttress them as "subjectivist". Let me use Josef Cardinal Ratzinger as an example. In describing proportionalism, Ratzinger states: "the morality of an act depends on the evaluation and comparison made by man among the goods which are at stake. Once again, it is an individual calculation, this time of the 'proportion' between good and evil" ([20], p. 90). Ratzinger has nothing against a calculation of goods and evils, as he makes explicitly clear. What bothers him is that he sees contemporary forms of this as rooted in "the 'reason' of each individual." He contrasts this with a morality based on revelation.

Two brief remarks are in place. First, Ratzinger implies that if a morality is revealed, no personal evaluation is necessary. That means that God's revelation, what Ratzinger calls His "instructions for use", is so utterly detailed that it covers all imaginable variations and conflicts and dispenses with human reflection. That is, of course, absurd and no one ever held it.

Second, if Ratzinger's real concern is the *individual* (hence potentially subjectivist) character of the discernment to be made, then the proper reply is twofold. First, evaluation by an individual does not mean individualistic evaluation. We form our consciences in community. Second, as Edward Vacek, S. J., has noted, being true to the relational character of reality is not being arbitrary and subjectivist ([24], p. 296).

(h) Complementary vs Contradictory Pluralism

Some have attempted to enlighten the discussion by the use of this distinction. Let me use Thomas Dubay as an example ([6], pp. 482–506). Clearly Dubay sees contradictory pluralism as inconsistent with scriptural insistence on unity, destructive of practical pastoral guidance, and deadening to the Church's commission to speak out authoritatively on important moral matters.

But when is pluralism "contradictory"? Dubay's answer: when it touches "important moral and disciplinary matters." We must have unity based on a "secure knowledge of the moral implications of many acts."

Here I believe we must ask, what are these "*important* moral and disciplinary matters," what are these "*basic* matters or norms" confused by a contradictory pluralism? Are they rather detailed and concrete conclusions representing the application of more general norms? Or are they the more general norms themselves? Dubay's terminology ("basic matters or norms") suggests the latter, but I suspect he is really looking for unity and security at the level of application; for he speaks of "a secure knowledge of the moral implications *of many acts* . . . ." So, how basic is basic?

The point I am making is that if discussions of pluralism are to be enlightened by the distinction between complementary (acceptable)-contradictory (unacceptable) pluralism, the implied criterion of *basic* must be spelled out. The same can be said of the usage "legitimate" pluralism (acceptable) and "radical" pluralism (unacceptable).

A personal reflection. A past tradition easily led us to believe that "basic" had to do with matters of self-stimulation for sperm-testing, removal of ectopic fetuses, actions that are *per se graviter excitantes*, co-operation in contraception and a host of very concrete applications. We felt we ought to possess and did possess a kind of certainty and subsequent security in these matters, and that our certainty was founded on the natural law. These, I submit, are not "basic matters or norms", if by this term is meant material on which we must agree if our Christian unity is to remain integral. There is plenty of room for doubt, hesitation, and change, even contradictory pluralism (dissent) at this level of moral discourse.

And yet, because the magisterium did get involved in such detailed practical applications in the past (e.g., allocutions of Pius XII, responses of the Holy Office), and in a way that was authoritative, it gave credence to the notion that our moral unity is or ought to be located at this level, and that disagreement or pluralism at this level is a threat to unity. In my view, that is an unrealistic view of both unity and the capacity of the human mind for certainty.

This is, of course, a key point, perhaps *the* key point, in discussing pluralism in Catholic moral theology. It must suffice here to note that my judgment is not altered by an escalation in the type of argument or analysis sometimes used to establish positions. For instance, Carlo Caffara has argued that contraceptive interventions contravene the rights of God and prevent God "from being God" ([8], pp. 363–382). John Paul II has made similar statements. Such statements do not thereby show that the issue of contraception is so basic that pluralism on the matter is ruled out as destructive of Christian unity. For one thing, these are analytical theological arguments, and their validity is subject to rigorous theological critique. For another, nearly every moral conclusion can be cast in such broad theological terms that it seems to involve the assertion of divine governance, divine providence, trust in God, etc. This destroys the distinction between basic and non-basic in principle. Furthermore, it offends common sense.

IV. AREAS OF PLURALISM IN MEDICAL ETHICS

When I speak of pluralism here, I mean it in a narrow and technical sense. In such a sense it comprises differences between *established and recognized theologians* (and philosophers). These differences fall into two categories: concrete and general.

(a) Concrete

Issue areas where there is factual pluralism are the following:
– Contraception, sterilization.

- Reproductive technologies (artificial insemination, *in vitro* fertilization, masturbation).
- Abortion and the status of the pre-embryo.
- Life preservation (newborns; nutrition-hydration question).

(b) General

The following points seem to cover the most disputed areas:
- Methodology (in the understanding of normative statements).
- Teaching authority (e.g., the significance of magisterial statements; the 1971 *Directives for Catholic Health Care Facilities*).
- Public policy (e.g., Medicaid and abortion).
- Hospital practice (e.g., cooperation; joint ventures).

V. PERSONAL PROCEDURAL SUGGESTIONS

These final reflections root in a twofold conviction. First, it would be unwise to attempt to "solve" (i.e., do away with) the problem of pluralism by the use of authority. History (*Tuas libenter, Pascendi, Humani generis*) shows that while such interventions may achieve a short-term "peace of a kind" ([1], p. 299), the long-term fallout is detrimental. This point was recently underscored by Archbishop Rembert Weakland[25]. Weakland referred to the first decade of this century as a time of "theological suppression" and added that it resulted "in a total lack of theological creativity in the U. S. A. for half a century." Weakland referred to a "better way of proceeding" and cited John XXIII's opening speech at the Second Vatican Council where the pope stated that the Church "prefers to make use of the medicine of mercy rather than that of severity." John XXIII added: "She considers that she meets the needs of the present day by demonstrating the validity of her teaching rather than by condemnations."

Second, it is impossible to give criteria, except of the most general (and therefore not very useful) kind for determining the limits of acceptable pluralism. For these reasons I want to offer some procedural suggestions for living and dealing with pluralism. These suggestions touch the magisterium, theologians (and scholars generally), and the Catholic public.

Magisterium

It might seem arrogant for a theologian to suggest to the magisterium how it ought to conduct itself. In reality, however, it is not. What the magisterium teaches, how it teaches, with what authority and consultative processes are properly *theological* questions. Furthermore, teaching is never done in a vacuum. It is done in a historical moment and the taught live in particular and varying cultures. Theologians and other scholars who reside in those cultures should be presumed to have some idea of what effective teaching is in their particular cultures. With that in mind, I offer the following suggestions that touch the moral teaching of the magisterium.

– The magisterium should formulate its teaching in an *open, revisionary* teaching process. This means, among other things,that the consultative process is not narrowed to draw on those only who agree with a foreordained position.

– The magisterium must be *well informed* scientifically. Once again, this means broad consultation of all competences. One sees and praises this in the *Declaration on Euthanasia* but regrets its obvious absence in the *Declaration on Certain Sexual Questions*.

– The utterances of the magisterium should be appropriately *tentative*. Excessive claims, apart from their theological inaccuracy, are a source of future embarrassment to and confusion in the Church.

– Whenever feasible, the magisterium should be explicit about the *ecclesial status* of a teaching (what used to be known as the "theological note"). Karl Rahner argued for this for years and the American bishops followed the suggestion in their pastoral *The Challenge of Peace*.

– Moral teaching should be bolstered by *persuasive analysis*. It is counterproductive in our time to urge conclusions solely with the weight of formal authority. This is particularly true of condemnations. Practically, this means that the magisterium will not authoritately propose conclusions against a strong theological counterposition.

– Persons in authority should be *realistic*. I mean to suggest under this rubric that the magisterium should not expect agreement with every official statement. So-called "official teaching" can be in various stages of development, as Ladislaus Orsy, S. J., has shown ([13], pp. 396–399). Furthermore, doctrinal development cannot be excluded, a point emphasized by Archbishop John Quinn in the synod of 1980 ([18], pp. 263–267).

– Finally, if we are to live with pluralism in a joyful and mutually supportive spirit, it is essential that there be in place truly fair procedures for the implementation of fraternal correction. "Due process" must mean more than "what Rome thinks due." The Church can and should, without apology or embarrassment, learn a lesson from the western democracies in this area. Until it does, we shall continue to hear even very legitimate interventions descibed in star-chamber language. We shall hear references to the fact that [in the Curran case] "Ratzinger was prosecutor, judge, jury and executioner"[9]. Greeley concludes: "It doesn't work anymore. The leadership of the American church knows it doesn't work, knows it is counterproductive, knows that burning of research data and star chamber trials are disgusting and terrifying phenomena to most American Catholics." We may fault the language and deplore the bile. But the substantial point survives such faulting and deploring: to the American Catholic certain methods can be far more "confusing" than any pluralism in moral theology.

Theologians

– Theologians must present their analyses and conclusions with *due modesty*. They should not claim practical probability for a position if there is no or little theological support for it. The claim to prophecy must be rare and reluctant.

– Theologians must be *pastorally sensitive*. Specifically, they must realize that it is sometimes necessary for the magisterium to present an authoritative position, even if it is tentative and temporary.

– Theologians – at least some – must renew and fortify their resolve to present fairly (in the best light) positions with which they disagree. This means practically the avoidance of "isms", accusations of harm to the Church, *post hoc ergo propter hoc* allegations, etc.

– The traditional doctrine of the "presumption of truth" favoring the magisterium must play its appropriate role in theological discussion, notwithstanding the fact that it is *only* a presumption and that such a presumption can be undermined in practice by a shortcircuiting of the deliberative process.

– Theologians must write, speak, and act in a way that fosters respect for the magisterium. This is, to a Catholic, self-evident, even if at times some members of the magisterium make the task more arduous than it ought to be.

– Finally, it should go without saying but will not, that theologians and scholars should be ready, willing, and able to admit mistakes – especially to each other. Without such readiness, differences degenerate into distance, and ultimately disorder.

Catholic Public

The suggestions I make here are perhaps the most difficult to realize in practice. But that is not an excuse for not trying.

– The Catholic public must have a much more accurate notion of the place of the magisterium and scholarship in the Church. The gigantic character of the educational task becomes apparent when we recall that many priests have not achieved this accuracy. Far too frequently, even priests conceptualize the relationship competitively, in terms of the magisterium *or* scholars.Furthermore, the notion of infallibility is so badly misunderstood by the public that one nearly despairs of rectifying it.

– The public must be educated to understand that the Church does not have all the answers, or an immediate one to a new question. Vatican II acknowledged this openly and realistically. "The Church guards the heritage of God's word and draws from it religious and moral principles, without always having at hand the solution to particular problems"([1], pp. 231–232).

– The Catholic public – and most especially, of course, competent professionals – must be educated to the responsibilities inseparable from their own competence. This was also clearly stated by the council fathers([1], p. 244).

– The Catholic public must be educated more than they are to two facts: (a) priests do not always speak accurately and critically on moral issues. (b) Moral theologians, in their public statements, do not always speak for the Church, or even for all moral theologians. On occasion, they present their own opinions which can vary all the way from "strictly personal" to "Church teaching". Somehow or other we must find a way to clarify this and the burden of doing so rests, I believe, largely on theologians and scholars.

– Finally, the Catholic public must be educated to the fact that the media thrive on confrontation. Many issues are not nearly as controversial as they are sometimes made to appear.

There will always be tension and conflict on moral questions in the

Church. Where there is no pluralism in that sense, there probably is very little creative thought going on. And when that happens, problems much deeper than those associated with pluralism will begin to appear – sooner or later, soon at the latest. That is why the temptation of easy solutions is precisely that, a temptation.

University of Notre Dame
Notre Dame, Indiana
U. S. A.

BIBLIOGRAPHY

1. Abbott, W. M., S. J. (ed.): 1966, *Documents of Vatican II*, America Press, New York.
2. Bernardin, Joseph Cardinal: 1983–4, 'Cardinal Bernardin's Call for a Consistent Ethic of Life', *Origins* **13**, 491–494.
3. Bernardin, Joseph Cardinal: 1983–4, 'Enlarging the Dialogue on a Consistent Ethic of Life', *Origins* **13**, 707–709.
4. Connery, J. R., S. J.: 1986, 'Quality of Life' *Linacre Quarterly* **53**, 49–53.
5. Donavon, D. L.: 1985, 'Church and Theology in the Modernist Crisis', *Proceedings of the 40th Annual Convention* (CTSA), 40.
6. Dubay, T., S.M.: 1974, 'The State of Moral Theology', *Theological Studies* **35**, 482–506.
7. Ford, J. C., S. J. and Germain Grisez: 1978, 'Contraception and Infallibility', *Theological Studies* **39**, 258–312.
8. Fuchs, J., S.J.: 1984, 'Das Gottesbild und die Moral innerweltlichen Handelns', *Stimmen der Zeit* **202**, 363–382.
9. Greeley, A.: 1986, 'Red Baron of the Vatican Strikes Again', *Chicago Sun Times* (September 14), C2.
10. Grisez, G. and J. Grisez: 1986, 'Theology Within "the certain gift of truth"', *National Catholic Reporter* (July 4), p. 14.
11. Hannan, Archbishop Philip: 1986, N. C. News Release (March).
12. Kaiser, R. B. 1985, *The Politics of Sex and Religion*, Leaven Press, Kansas City.
13. Orsy, L., S.J.: 1986, 'Reflections on the Text of a Canon', *America* **154**, 396–399.
14. Petre, M.: 1918, *Modernism: Its Failures and its Fruits*, T. C. and E. C. Jack, London.
15. Pinckaers, S., O.P.: 1982, 'La Question des actes intrinsiquèment mauvais et le "proportionisme"', *Revue Thomiste* **82**, 181–212.
16. Pope John Paul II: 1984, *Reconciliation and Penance*, USCC, Washington, D. C.
17. Quay, P., S. J.: 1975, 'Morality by Calculation of Values', *Theology Digest* **23**, 347–364.
18. Quinn, Archbishop John: 1980, 'New Context for Contraception Teaching', *Origins* **10**, 263–267.
19. Rahner, K.: 1976, 'Basic Observations on the Subject of Changeable and Unchangeable Factors in the Church', *Theological Investigations* **14**, Seabury, New York, pp. 3–23.

20. Ratzinger, Joseph: 1985, *The Ratzinger Report*, Ignatius, San Francisco.
21. *Schema constitutionis pastoralis de ecclesia in mundo huius temporis: Expensio modorum partis secundae*: 1965, Vatican Press, Rome.
22. Schüller, B., S. J.: 1984, 'Die Quellen de Mortalität', *Theologie und Philosophie* **59**, 535–559.
23. Schüller, B., S. J.: 1986, *Wholly Human*, Georgetown University Press, Washington, D. C.
24. Vacek, E., S. J.: 1985, 'Proportionalism: One View of the Debate', *Theological Studies* **46**, 287–314.
25. Weakland, Archbishop Rembert: 1986, 'The Price of Orthodoxy', *Catholic Herald* (September 11 and 18), 1–3.

JUDE P. DOUGHERTY

ONE CHURCH, PLURAL THEOLOGIES

At any earlier time in this century, pluralism referred to a religious pluralism in which the Church was regarded as one sect among many. Today, the meaning of pluralism has been expanded to include irreligion as well as religion. In fact, the pluralist society in which the Church vies for allegiance is largely a secular society. This is true of the West in general, and it probably became true in Europe before it became so in the United States. How it became so in the States is a long story that need not be told here.

To confine attention momentarily to religious pluralism in the United States, one can identify at least 215 religious bodies with some claim to national recognition. That number does not include the countless storefront or one-of-a-kind churches that are to be found in both urban and rural areas. Historically considered, the Protestant emphasis on personal judgment, choice, and commitment, and on individual response to the teachings of Christ, has resulted in the loss of a commonly accepted code of doctrine and morality. The Catholic Church with its emphasis on the magisterium and on tradition has fared better. Possessing a unity and a conviction not found in Protestantism, its problem has been different. Since the Reformation, where it has existed as one faith among others, the Church has had to reconcile two demands: (1) the need to show that Christian revelation as mediated by the Church discloses truth about religion and about God and (2) the need to show how other people who are not Catholic can still be involved in valid religious activity and thought. From the Catholic point of view, the challenge is to reconcile truth with toleration and with the pluralism mentioned. The danger in this area is that of allowing the Catholic position to become merely one more opinion among many and not the definitive truth about God and religion.

Now attention is called to a pluralism within the Church itself, a pluralism in the theological order which presumably makes it difficult for the Church to teach with a united voice. But need we be alarmed? Theological pluralism, I am convinced, is the normal state of affairs, and even divergent opinion in the realm of morality is to be expected. When

169

Edmund D. Pellegrino et al. (eds.), Catholic Perspectives on Medical Morals, pp. 169–172.
© *1989 Kluwer Academic Publishers. Printed in the Netherlands.*

a distinction is made between moral norms and the prudential application of those norms, it is immediately apparent that people who share the same code of moral principles may disagree about their relevance to a given situation. Acknowledging this is not to deny that the Church faces a different sort of question, namely, the role of the theologian, the intellectual, if you will, in the Church. This problem is not exclusive to the Church but is found in other collectives as well. Hans-Georg Gadamer has addressed the problem from the vantage point of his hermeneutic methodology. The larger problem of the expert in society may be put in this way.

Does the control of a group (of whatever kind) belong as a right to the few (the experts) exclusively and not to the many? Or are the many entitled to share the control, because the limited knowledge of the many, when it is pooled and critically restated through mutual discussion, provides a lay consensus capable of revealing certain of the limitations of interests in the expert's point of view? Or thirdly, may it be held that this consensus knowledge of the many entitles them to have full control, excluding the experts.

This manner of questioning does not solve our problem but it does suggest that questions regarding the role of the expert within the Church is something less than unique. Most of the time intellectuals are in battle with the rest of the community, and this no less in the Church than elsewhere.

To return to an earlier point, pluralism in theology should not be confused with pluralism in the Church. Theologies, like philosophies, of their very nature are plural. Some theologies are useful to the believer, some are not. The work of the theologian resides not in any practical undertaking but simply in making discoveries. It is not for the theologian to decide whether he has uncovered anything important or has succeeded in capturing the mind of the Church. He will be judged from without. While some theologians make an attempt to be of use to the Church, others do not. Nothing requires a theologian to place his intellect in the service of the Church. That he does so is the result of a free act. If he does so, then he must avoid certain traps. For one thing, he cannot be parochial either in time or place. For another, the theologian who would be of service must have a proper grasp of his role in the Church. He is not an official teacher. He is a scholar whose writing and discourse are governed by professional canons of behavior, but that does not guarantee orthodoxy. By the very nature of his enterprise the theologian is likely to take a thesis and push it to its limits. He can easily overstate his case or exaggerate a point of view. The bishop, on the other hand, is chosen for his prudence after a very careful selection process. A theologian publicly engaging his bishop or the Supreme Pontiff has lost both his modesty and his sense of professional decorum. Bishops and theologians are not peers.

No one is scandalized by plural theologies; rather, the scandal lies with theologians purporting to teach over the heads of official teachers. The existence of many Protestant theologies is a fact of intellectual life and no student of religion is surprised by that. Theologies define different Protestant bodies. The bodies themselves are created by the theologies. Within most Protestant bodies the authority of the theologian is not circumscribed by a magisterium. As teachers, neither the theologian nor the ecclesiastical office holder is privileged. But such is not the case with the Roman Catholic Church. In spite of plural theologies from Origen through Augustine, Bernard, Anselm, Bonaventure, Aquinas, Scotus, Campanella, and Suarez, and in our own day, de Chardin, Rahner, and Lonergan, the Church in a significant way has been able to speak *sole voce* because of its hierarchial structure.

Without the authoritative influence of the Church, all too fallible intellect can go astray. Perhaps no one was more aware of this than the great Origen who wrote: "I want to be a man of the Church. I do not want to be called by the name of some founder of a heresy, but by the name of Christ, and to bear that name which is blessed on the earth, it is my desire, in deed as in spirit, both to be and to be called a Christian. If I who seem to be your right hand and am called a presbyter and seem to preach the word of God, if I do something against the discipline of the Church and the rule of the Gospel so that I become a scandal to you, the Church, then may the whole of the Church, in unanimous resolve, cut me, its right hand, off, and throw me away."

Many contemporary presentations fail not so much in substance but in tone. When Catholicism, indeed all Christianity, is under serious attack from without, it does not make sense to undermine legitimate authority, particularly when it is providing normal leadership for a people who look to it for guidance. When John Paul II states that there are "intrinsically illicit acts independent of the circumstances," most hearers will know what he is talking about. They do not need to be told that there are no acts in the abstract, that all acts occur within a set of circumstances and that sometimes the way we denominate acts includes the circumstances in which a particular act occurs. Most readers will understand that what the Holy Father has in mind, for example, is the judgment that no external circumstances will justify the direct and intentional aborting of the unborn; not the health of the mother, the poverty of the family, or the possibility of a physical defect in the child. Similarly, it serves no useful purpose to distort the remarks of a Cardinal Ratzinger who in arguing against subjectivism is clearly not denying the necessity of personal evaluation. Ratzinger is merely affirming that personal evaluation is to take place against objective canons.

Bishop-bashing may be a normal clerical parlor-sport, but publicly to misconstrue or place the worst possible interpretation on ecclesiastical pronouncements is, I think, unprofessional. No pluralism justifies a destructive attitude toward authority in one's community. If that authority speaks ambiguously, then the task is to sort things out in the light of the intended message. There is nothing in the nature of the discipline requiring the theologian to be critical.

A final point. The Church has not invented morality. It is to her credit that in every period of her history she has adopted the highest moral principles available. It knows that weak men must fail. But while it provides sacraments for his reconciliation, it has never given him excuses for his failure. Children demand a stern father and respect one when they get one. It does not advance the cause of parental authority to say that Pop may be wrong or too severe or precipitous. We all know that emphasis can be misplaced, but those who respect the Church know that the life to which it calls us is a life of self-fulfillment and that whatever its shortcomings it is never wrong in directing us to become the most that we can be.

Catholic University of America
Washington, D.C., U.S.A.

GERARD J. HUGHES, S. J.

IS ETHICS ONE OR MANY?

INTRODUCTION

A common theme runs through this entire paper, and that is the
phenomenon of moral disagreement. I say "phenomenon" in order not
to commit myself in advance to any view about whether or when moral
disagreement is genuine or merely apparent: indeed, as will appear,
whether genuine moral disagreement is even possible is one of the most
important issues at stake. But even those who concede that genuine
moral disagreements are possible have different views about the precise
nature of the disagreements that there are, and about the best way of
trying to resolve them. The various "pluralisms" in morality derive, I
think, from the variety of analyses given of the phenomenon of moral
disagreement. I believe that these different pluralisms raise very different issues which ought to be kept separate. I shall be arguing that
in three, but not in all, senses of "pluralism" a pluralist position in morals is preferable.

I. NON-CONTROVERSIAL IGNORANCE

The first is the kind of pluralism resulting from our ignorance of those
non-moral features of a situation on which depend our moral judgments
about what ought to be done. Thus there is some controversy at the
moment about how one ought to treat patients suffering from arthritis,
or whether babies should be given aspirin when suffering from in-
fluenza, since there is dispute about the side-effects that might be
expected from the various drugs able to be employed. Again, since the
precise nature of some illnesses, for instance epilepsy or schizophrenia,
is not entirely understood, there are several views about the best way in
which patients suffering from these illnesses ought to be helped. Again,
we might be unclear on non-medical grounds how far a patient would be
able to profit from a particular form of treatment. In a case receiving
some publicity in Britain, a hospital had a severely limited number of
home kidney dialysis machines and decided that a particular patient was
not sufficiently organized and persevering to make proper use of it – a

173

Edmund D. Pellegrino et al. (eds.), Catholic Perspectives on Medical Morals, pp. 173–196.
© *1989 Kluwer Academic Publishers. Printed in the Netherlands.*

view that was hotly contested not merely by the patient himself but also by social workers who knew him well. Other examples of moral problems where we have insufficient knowledge to be able to arrive at clear solutions might include decisions about the schooling of children known to be carriers of the AIDS virus; and the possible genetic effects of large hormone dosage on a woman's ova, and implications of this for various methods of helping infertile women to conceive. In each of these examples, conclusive evidence of various kinds is lacking, and moral decisions have to be made in situations where various conflicting moral choices can legitimately be defended on the basis of the knowledge actually available. I take it that there is no difficulty in admitting that there can be a plurality of defensible moral views in such cases. Two general points are worth making briefly even about these uncontroversial cases, however. The first is that there seems to be no difficulty in such cases in admitting that our moral decisions must wait on our factual knowledge of human beings and their environment, since without that knowledge it is not clear what it is that we are actually *doing*; and hence there seems to be little difficulty in admitting that our moral knowledge – our knowledge of how we should act – is in many ways deficient. I say this because it is sometimes assumed that we – whether moral philosophers, or the medical profession, or the Church – must somehow always be in possession of the answer, if only we looked carefully, or honestly, enough to find it. Second, I would note that the mere fact of disagreement in such cases does not of itself threaten the claim that there is a right answer even if, as things stand, we have no way of knowing what that answer might be. Sometimes the right answer becomes obvious with time. Thus with hindsight we can now see that doctors who prescribed thalidomide for pregnant mothers acted wrongly, even if their moral decision about the best mode of treatment might have been perfectly defensible on the evidence available to them at the time. Doubtless, decisions taken by doctors today in equally good faith will in thirty years time be seen to have been equally wrong. In the meanwhile, it may and usually will be the case that in many instances several incompatible views can all be rationally defended. And that is one kind of pluralism in ethics.

Other differences of opinion involving ignorance are more complex because further and larger issues are involved. What is fundamentally in dispute may be whether or not certain information would even be relevant to a moral decision. A person who thinks the information irrelevant may know nothing of the information in question, and may have a perfectly clear moral view as a result; another, who takes the information to be highly relevant, might find himself in moral perplexity precisely because his knowledge of that information is incomplete. I suggest that even where the relevance of the alleged facts of the matter is in dispute, it would still be of enormous help first to try to be clear about just what the facts of the matter really are. Thus, prior to the dispute about whether couples should be enabled to have children independently of sexual intercourse, we ought to try to discover to what extent, if any, the relationships between the couple themselves and between them and their children are affected by the fact that children have been conceived by different methods. Prior to a discussion about the proper limits of doctor-patient confidentiality, it would surely help to know as much as possible about the actual effects of various policies. The same goes for many other spheres of moral conduct – notably sexual ethics, public economic policy, defense policies – which are not directly the concern of this conference.

There are two main reasons for saying that one ought to be in favor of trying to discover such information, even if one does not believe it to be relevant to any moral decision. The first is that the person who believed it to be relevant might, in the light of what is discovered, reach the same conclusion as the person who believed it to be irrelevant, albeit for different reasons. And practical agreement is worth having. The second reason is that the person who thinks the information irrelevant might come to think differently once the information is actually in front of him. It is easy to dismiss as irrelevant information one does not have, and questions of relevance, as we shall see, are not so philosophically open and shut that one can afford to run that risk.

The major problem with this suggestion is that it may be that the only feasible way of finding out such information might be by engaging in conduct (at least in default of the information) one believes to be wrong. One cannot determine the risks involved in genetic research without engaging in it, nor the effects of giving contraceptives to minors unless they are given. I do not have any very satisfactory solution to offer. On the other hand, given the variety of moral policies that are in fact pursued, it may often be the case that such information could be discovered in any case. Where it can, I think every effort should be made to ensure that it is as widely known as possible.

In sum, it is obvious that moral decisions should be made on the basis of the best relevant information available. I suggest that even when the relevance of the information is not clear, the information should still be sought. Uncontroversially, however, a plurality of moral views can be expected to be rationally defensible where relevant information is lacking.

II. THEORETICAL DISPUTES

A second kind of pluralism derives from the fact that there are unresolved theoretical disputes of a more philosophical character which have a bearing on moral decisions. These are of quite a different nature from the disputes arising from ignorance which we have so far considered. In the standard cases of ignorance, there was nevertheless agreement on what it was that we needed to know in order to make good moral decisions, and there was agreement about how such knowledge might in principle be obtained. In contrast, the theoretical disputes are precisely disputes about the method by which progress is to be made towards better moral decisions; and it is even disputed whether there is any truth of the matter to be discovered. I shall confine myself to two large areas of difficulty that are of especial relevance to medical ethics – the nature of the human person, and the proper pattern of moral argument.

1. The Status of the Human Person

Even the most cursory examination of disputes about the human person will reveal that they have been bedevilled by equivocation and by failure to see precisely how the arguments of the other sides in fact worked. This is particularly evident in connection with disagreements over abortion, in-vitro fertilization, and genetic research, though it is also to some extent true at the other end of the life cycle as well. For instance, the word "human" has been ambiguously used. There is little, if any, disagreement about the facts of embryology, nor about whether a fetus is human rather than, say, bovine or equine. There are well-recognized procedures for settling any disagreement that there might be on any such issue. But if "human" is used to mean "human person", then of course there is dispute about whether a fetus that is admittedly human in the first sense is or is not a human person. Similarly with "soul"; if to have a soul is the same as to be alive, then of course there is no dispute about a fetus having a soul from the first moment of conception. But if to have a soul is taken to entail being a person with human rights, then of course there is a dispute whether a human fetus has a soul in this sense. Obvious as these points are, it is nevertheless the case that many arguments and counter-arguments have simply failed to meet one another because such elementary equivocations were not attended to. Once they are, it rapidly becomes obvious that the apparent point of disagreement is not the basic point of disagreement at all. Similarly with the use of "person". Some radical feminist groups would insist that the embryo is simply a part of the mother's body, since once that position is accepted the moral conclusions they wish to support will more easily follow. Similarly, Catholics have at least in recent times adopted the view that the embryo is a human person from the first moment of conception, since that position gives greater support to a particular moral view about the wrongness of abortion and the wrongness of certain procedures connected with IVF and of certain aspects of genetic research. On both sides, the concept of "person" is morally charged. In contrast, many philosophical articles on personal identity have been written quite independently of any such moral considerations. Theology, psychology, sociology, and the law all operate with yet other notions of "person". In the midst of such equivocation, it is no wonder that argument has proved impossible, and that misunderstanding and anger have been the only result.

Similar considerations apply to the problems involved in deciding which criteria to adopt in deciding whether or not somebody is dead. There need be no dispute about the purely medical facts – about the irreversibility of coma, or the absence of activity in the brain-stem, and so forth. But the dispute about what these medical facts add up to remains, and there is no agreed procedure for settling it, nor indeed any agreement about what *kind* of question is being asked when one asks whether someone is dead or not.

It is, I suggest, the task of philosophy both to reveal such misunderstandings, and to make proposals about how our conceptual apparatus can be reorganized in a way that is both coherent and serves the purposes for which we have developed it. In so doing, I take philosophy to be pursuing the truth of the matter. As things stand, I do not believe that there is a concept of "person" that is both wholly coherent and seviceable. As a realist, I therefore believe that the dispute is far from being merely a verbal one, to be settled by appeal to normal usage, or simple stipulative definition, as one might, for instance, so decide whether or not a particular article of furniture was a table or a desk. For I take it that to be a human person is to be a member of a natural kind, and that as things stand we simply have no clear knowledge of what a person *is*. As a realist, I am committed to the view that there is a truth of the matter to be attained here; but as a realist I am also perfectly ready to admit that our grasp of this truth is still very inadequate.

In any case, this much seems clear to me: Any moral conclusions should depend on the application of a notion of "person" that has been arrived at by an independently-conducted philosophical debate, rather than function as premises in that debate. As for the philosophical debate itself, I am of the opinion that the demise of the more positivist versions of empiricism has at least opened the door to the possibility that argument can proceed on sounder lines than in the more recent past. In my own view, a more Aristotelian approach to the question does, in conjunction with our knowledge of embryology, hold out some hopes of at least a reasonably defensible philosophical position which would command a considerable degree of support. To discover what a human person is is both a philosophical enterprise and a matter of empirical research guided by that inquiry. The fundamental problems involved are not moral problems. As such, they are confused, not clarified, if moral considerations are allowed into the discussion at this stage. It is only when the philosophical and empirical issues have been clarified that we might on that basis proceed to discuss the moral consequences that might follow. In the meantime, we could with profit at least recognize the extent to which we have simply failed to understand the nature and roots of our disagreement.

What is fundamental to the disagreement between the traditional Catholic and the radical feminist over abortion is certainly not the status of the fetus; *that* dispute is but the symptom of quite other concerns on both sides. More progress might be made if these other concerns were identified and examined separately from the necessary philosophical discussion about what it is to be a person. Thus, for instance, one might seriously ask what rights something has which is admittedly a human person, and what the basis of those rights is. Not all persons here and now have the right to marry – the basis of that right does not exist in a five-year old, for instance. Is it the case that all persons here and now have the right to life? If so, what is the basis of that right? What considerations legitimately override the right to life if someone does possess it? To attempt to short-circuit such difficult problems simply by insisting on, or denying, the applicability of the word "person" or "potential person" is to misunderstand the nature of the problems involved. Both sides to the disputes about abortion, in vitro fertilization, genetic research, and similar issues have become progressively more irritated by what they take to be the failure of their opponents to take the real issues seriously. Both sides, as it seems to me, have contributed to this irritation by attempting to reduce difficult, and possibly threatening, issues to simple problems which can be settled by verbal intransigence and by encouraging the suspicion that the real reasons for their positions are not in fact those that are openly discussed.

The fact remains, however, that on the substantive issues concerning the nature of the human person and the relationship between being a person and having rights there is an insufficient degree of agreement for us to hope that once the problems have been clarified they will disappear. I believe it to be only honest to admit that here too, at the philosophical level, we are to some extent ignorant of the truth. From which it follows that different positions can be rationally defended. That is another reason for pluralism, this time of a theoretical kind.

2. The Pattern of Moral Argument

Similar problems beset discussion of the nature of morality itself, and the proper form for acceptable moral arguments. The simplest way to illustrate this is to ask whether it is true that circumstances alter cases. Well, obviously, circumstances alter cases. Different patients will have to be treated in different ways. And even the same patient might have to be treated in different ways depending on the resources available, for instance, or on the existence of legislation stipulating which drugs may or may not be prescribed. A clinic in Ethiopia might have to treat patients in ways and under conditions that would simply not be countenanced in London or Washington. To the extent that circumstances alter cases, patients with the same complaint will have to be differently treated. The examples are uncontroversial because, so far as I know, these difficulties would receive the same solutions at least in most moral codes. But *which* circumstances alter cases and why is, of course, much more problematic, since it is related to very general questions about what evidence is relevant for establishing moral conclusions. I have argued elsewhere([6], ch. III) that questions about the relevance and bearing of non-moral facts on our moral judgments have frequently been obscured by being approached from other, to my mind less profitable, directions. I will try to summarize my views briefly here, insofar as they are directly relevant to my present concerns.

Firstly, the debate about whether actions are right or wrong in themselves or because of their consequences is, to my mind, largely misplaced, for two main reasons. First, there is a confusion between types of action (which as such do not have consequences at all) and individual actions (which do). And second, if the question is asked whether an individual action is right/wrong in itself or because of its consequences, one must first identify which is the action under discussion. Many different criteria for identification have, historically, been given: for Bentham, the action itself would be the mere voluntary bodily movement – and on this definition, of course the consequences (and circumstances) will obviously be decisive, and the action in itself will be morally neutral. For Kant, the action will consist of the agent's behavior as expressed in the agent's maxim – a definition which expresses his view that what is relevant to specifically moral appraisal in the first instance is the will of the agent rather than his behavior as such. Other moralists have drawn the line between the action itself and its consequences in different places in ways reflecting a moral judgment they have already made about the relevance or irrelevance of the various features of the agent's behavior. Thus, what an agent does could be truly described as moving his hands, as performing an operation, as rendering a woman sterile, as implementing a means of family planning, as removing an important source of emotional stress, and so on. We will doubtless have views about the rightness or wrongness of each of those *types* of action. But that does not help us to settle the prior question, how the agent's behavior is to be classified in terms of those types. Nor does it help to settle the further question, how we should evaluate a piece of behavior which is an instance of two types of action, one morally right and the other morally wrong. Dispute about this is a dispute about what is morally relevant, and rival answers will have to justify their views by appeal to a wider thesis about moral relevance in general.

Similar considerations apply to questions about whether there are any absolute moral principles, or whether there is any action which is absolutely right or wrong. "Absolutely" here is far from being a clear term: it is often taken simultaneously to be the antonym of "relatively", to be a synonym of "in itself", and to mean "in all circumstances", even though the three senses are quite different, and do not even entail one another. And the issue is further confused by the point I have already made about the ambiguity of the term "action". The traditional terminology has often had the effect of obscuring the real nature of the disagreements involved.[1] The same might be said of arguments about the principle of Double Effect. I think it has been conclusively shown ([2], pp. 1–16) that disputes about cases involving the principle, and about the best interpretation of the principle itself, are often confused because they use the word "action" in different senses, and because what is really at stake is a disagreement about which features of the situation ought to be taken into consideration when evaluating actions. Even were these terminological difficulties to be resolved in such a way that the real issues emerged clearly, it would still be true that my simple initial examples (of circumstances altering cases) were uncontroversial only because there is agreement on which features of the situation should have a crucial bearing on the moral judgment, and on which circumstances ought to make a difference.

Examples of cases in which this agreement is lacking are legion. I pick some simply because they have recently been in the news. How far is it proper to insist that the doctor-patient relationship must be considered confidential? Even where the patient is a minor and is being treated without the parents' knowledge? In Britain, one root of the disagreement has been the differing estimates of the effects of various policies on doctor-patient relationships generally, and on the incidence of pregnancy in girls who are minors. Nobody is entirely sure, and to that extent the variety of opinion derives from ignorance. There is also a difference of opinion on whether this information (were it even available) would be relevant, given that confidentiality is involved. Rather untypically, some Catholics have found themselves in the unusual position of appealing to the (alleged) consequences for the child's welfare against some doctors who maintain that a breach of confidentiality is absolutely wrong. Again, in some of the discussion on the Warnock Report, it has been urged that the fact that the implantation of an already fertilized ovum bypasses the normal method of fertilization by intercourse is an argument telling against the morality of the procedure. Others simply do not see that this is a relevant consideration at all. Or again, in the case of severely handicapped neonates, much of the dispute seems to me to turn on which facts (or presumed facts) are even relevant: the quality of life which the child can on the medical evidence reasonably expect? The attitudes of the parents? The life-expectancy of the child? The prospects for proper care of the patient in later life when parents may be dead? The difference between killing and allowing to die?

In contrast to the simple cases with which I began, pluralism in these latter cases is commonly perceived as somehow problematic. Neither doctors, nor patients, nor society at large will simply say, "There are several defensible positions here, and we just are not in a position to say with certainty which one is correct," as they might very well in the simpler cases. Nor, in contrast to agreed instances where circumstances make a difference, is it enough to say "Although in other circumstances I would not do this, here of course the circumstances are different"; for their opponents would not accept that these circumstances should make any difference. They will assimilate such cases to other, clear cases; and it is precisely the legitimacy of these comparisons that is in dispute.

In contrast to the simple cases of ignorance and variation in circumstances where a plurality of moral responses is not perceived as threatening, pluralism is seen as threatening in the cases of basic philosophical disagreement precisely because there is no agreed framework within which the disputes are to be resolved. It is the framework of argument which itself is in question. The framework is critically important, for it determines the answers to such questions as "Which considerations are relevant?", "How does one begin to evaluate morally conflicting considerations?", "Under what moral categories should this behavior be classified and thereby assessed?", as well as more clearly normative questions such as "What kind of life is worth having?", "What rights does someone have?", "Is this really an act of kindness?" It is therefore quite unsurprising if there is a strongly adverse reaction to any pattern of moral discourse which appears to call that framework into question. And it is equally unsurprising if, in default of an agreed framework, moral disagreements become intractable. The net result is so often misunderstanding, suspicion, and acrimony, and the life of the moralist solitary, poor, nasty, brutish, and (at least in public) short. What I have said about moral philosophy could well be said also of theology, and of Christian Ethics, which is often in the unhappy position of enjoying the worst of both worlds.

I think that there are at least some reasons for optimism. The disagreements in moral philosophy are, of course, serious, but the re-emergence of moral realism as a viable option, the contemporary stress on the moral virtues as well as on moral principles, and recent work on the concepts of desire and need, seem to me to hold out some promise at least of agreement on the general basis for an acceptable moral theory. Most importantly, as I shall mention presently, recent intercultural studies of ethics might be expected to reveal rather more clearly what it is that is most fundamental in morality and which enables us to recognize the moralities of other cultures *as* moralities. I have already mentioned earlier in this paper some of the ways in which I think discussions about the rightness and wrongness of actions have been distorted by equivocation on the word "action". Many other such examples could be given. I hope it is not naively optimistic to suggest that progress can be made, and has been made in clarifying the real issues at stake. From a rather different standpoint, Garth Hallett's study of the tradition of Christian moral reasoning[4] is an example of the kind of approach necessary to understand the moral frameworks that have in fact been operative in Christian tradition and to see how they relate both to one another and to any approach we ourselves might, in the light of this understanding, wish to adopt. Quite apart from any theological considerations that might lead one to suppose that Christian tradition in this respect is worthy of our acceptance, it appears to me that a philosophical justification for at least broadly similar conclusions might well be forthcoming.

In short, I suggest that there is very reason to suppose that we can tackle just those issues about moral relevance which proved to be a major source of difficulty in the examples I quoted earlier in this paper. What I am at least clear about is that the problem is a problem about the nature of morality itself; and that it is to this, rather than to particular contentious issues, that our efforts should in the first place be directed. A plurality of views there may indeed be, in this as in so many other areas, but the convergence between the various possible theories seems to me a notable feature of moral philosophy in recent times. I venture to suggest that we are closer to discovering at least the core-elements which will be found in *any* account of what morality itself is, and how moral thinking of its nature must be conducted. Meanwhile, we must recognize the limits of our understanding, and the limited extent to which we have succeeded in formulating a moral view that is both coherent and satisfactory. To that extent, our grasp of moral truth is partial. The proper response to the recognition of our moral ignorance is twofold: the avoidance of dogmatism and an effort to ensure that the search for understanding is conducted in a climate of humility and tolerance – virtues that have all too frequently been conspicuous by their absence. But here, too, as with the previous philosophical issues connected with the human person, it seems to me that a plurality of views is rationally defensible, and that it would be less than honest not to admit that this is the case.

III. A MORE RADICAL PLURALISM?

In different ways, both kinds of pluralism for which I have so far argued have been argued for on the grounds that our grasp of what is needed to make and justify moral decisions is to a greater or lesser degree inadequate. I have, in short, appealed to two very different levels of ignorance. I now wish to argue for a pluralism in ethics which is at once more radical and more positive. I think the topic can best be approached via a discussion of moral relativism; partly because I wish to distinguish the position for which I shall be arguing from a relativist position; but more importantly because I think that an examination of the reasons for which relativism fails is of positive value for the point I wish to make.

A full discussion of moral relativism would be beyond the scope of this paper. I shall outline merely what I take to be the salient features.

1. Relativism – A Qualified Failure

Some alleged versions of moral relativism either rest on a confusion between the rightness of wrongness of actions and the blameworthiness of agents, or else amount to no more than the claim that if a person holds a moral belief, then he holds that belief to be true. Other attempted formulations, according to which a person's moral beliefs are supposed to be somehow constitutive of the truth of those beliefs have such implausible consequences that few if any modern relativists would be prepared to defend them. All these manifestly indefensible versions of moral relativism I shall simply pass over with just this brief mention.

The most defensible starting-point for a serious defense of relativism relies on the notion of a framework within which moral discourse is conducted. It will be argued that the truth of any moral belief is relative to the framework of moral discourse and culture within which that belief is held, since it is that cultural framework which gives the belief its content. It is open to this kind of relativist to claim that moral beliefs are true just when they correspond to the facts; but what the facts are depends on the way in which the framework interprets the world. The notion of a framework or "conceptual scheme" therefore needs some examination.

It is not even apparently relativist to maintain that different people hold different moral beliefs, or that moral beliefs which some people hold are equally held to be false by other people. Nor is it even apparently relativist to accept that the moral concepts of different cultural groups order the moral world differently – so the words *arete, eudaimonia, megaloprepeia* and many of the other virtue-terms in Aristotle do not stand in any straightforward relationship to moral terms in current English. To take an example in the other direction, I know of no obvious equivalent in, say, medieval Latin for our modern word "euthanasia". Questions of relativism arise only when the phenomena of cultural variation are related to some fundamental positions in epistemology and ontology. A Quinean scepticism about meanings and translation might, for instance, be used to give epistemological backing to the often-repeated view that the frameworks of different cultures are not merely historically conditioned (which is obvious) but also in some radical sense incommensurable. I shall not repeat the arguments of Quine or Kuhn here. I wish rather to draw attention to the implications such a position has for relativism about ethics. The view that truth in ethics is relative to the moral framework within which the moral utterance is made might be expressed in the following general form:

(R)   'M' is true in $(F)L$ if and only if P,

where 'M' is a sentence expressing a moral belief, L is the language in which 'M' occurs, and $(F)L$ is a framework of moral discourse available to speakers of L. As the roughly Tarskian form of (R) makes clear, (R) itself is expressed in a metalanguage, and makes an assertion in the metalanguage about 'M'. It is also assumed that 'P' is a metalanguage translation of 'M', and that 'P' is therefore expressed in the framework of the user of the metalanguage. To take an example, an instance of (R) might therefore be,

> 'Idina mwana akaa ni isha' is true in some framework used by speakers of Chaga if and only if female circumcision is permissible.

(The example is a real one. Not so long ago, it was revealed that African women were coming to London to have this operation performed by doctors in Britain.) The original relativist aim was to formulate the view that some moral belief might be true in one cultural framework and false in another. It is therefore essential that it be possible for the person asserting (R) to identify that moral belief. But if, on Quinean or other grounds, it turns out that there is no way in which we can justify translating "Idina mwana akaa ni isha" by "Female circumcision is permissible" rather than by anything else whatever, we will not be able to assign any one meaning to the native sentence, and hence we will not be able to identify which native belief it expresses. *Ad hominem*, one can then say that relativism of this radical kind is so radical that it fails to serve the purposes for which it has frequently been invoked. We can say neither that their morality differs from ours, nor that it is the same. Unless translation from their language into ours is guaranteed, then the principal relativist tenet is unstateable.

But now the relativist is in real difficulty. For if translation *is* assumed to be possible (on whatever grounds one takes Quine to be wrong), then the native language, and hence the moral beliefs expressible in that language, cannot be *radically* different from ours. In particular, if 'M' and 'P' have the same meaning, they must have the same truth conditions. If they did not have the same truth conditions, they would not have the same meaning, and hence there would be nothing surprising in the view that one can be true and the other not. The relativist is thus confronted with a dilemma; *either* the frameworks are so unrelated that we cannot identify the native beliefs at all, and hence cannot say whether or not they differ from ours: *or* translation is possible, in which case at least a radical relativism in ethics is false, for the moral belief expressed in Chaga would be true just when the corresponding belief expressed in English is true. Relativism therefore stands or falls on the question of whether it is in fact possible to identify a moral belief in a different framework as the same belief as one we might express within our own framework. The relativist, for the reasons I have just given, will be driven to claim that we cannot make such an identification, and that the beliefs stated in different frameworks must be radically incommensurable.

In fact, how *do* we identify the moral beliefs of others? Considerable attention has been given to this question for reasons which, I hope, will be obvious from what I have already said. That we do in general succeed in identifying the beliefs of others seems obvious enough. Everyday life would grind to a halt were this not the case. In more complicated situations, too, anthropologists will claim that they can, with patience,

understand the meaning of the statements and practices of other cultures, including their moral statements and practices. On what assumptions are such claims justifiable against the more sceptical forms of
relativism?

Richard Grandy, developing a suggestion of Donald Davidson,[2] has
argued for a "Principle of Humanity" ([3], pp. 439–452). We start by
assuming that other people have a pattern of beliefs and desires similar
to our own, formed, as ours are, by their interactions with the world.
We then try to make sense of their utterances and actions on the basis of
this assumption, modifying the details of our original ascription of
beliefs and desires as we go along, in order to make better sense of what
they in fact say and do. It is entirely within the spirit of this suggestion
that we might discover the moral thought patterns of some other culture
to be significantly different from our own. But we will still be using as a
benchmark our knowledge of the way our own desires and beliefs relate
to the world. And the test of our attribution of moral beliefs to others –
the test of our translation of their moral utterances – will be the accuracy
with which we predict their moral behavior in different sets of circumstances. Anthropological support for this general approach can be
found. Robin Horton, a philosopher specializing in West African
thought, argues that we will in fact distinguish between two levels,
which he terms primary theory and secondary theory[5]. Primary
theory involves such notions as "push/pull causation", the difference
between humans and other animals, the perception of time as before/
now/after; a relationship between space/time and causation, a world of
middle-sized solid objects, a distinction between self and others, a
system of perceptual differentiation. He argues that these will in fact
differ very little from one culture to another. There is also a level of
secondary theory which is characterized by the "hidden" explanations
of events in the primary realm. This will differ widely between cultures –
the Nigerians will explain all kinds of events in terms of spiritual
activities; the Westerner will explain them in terms of atoms and
molecules and sub-atomic particles, etc. But even here, there are
similarities between the two cultures: in both, the "hidden" entities are
frequently described in terms of models taken from the ordinary primary realm – spirits act somewhat as persons do, atoms can be thought of
as mini-solar systems, or waves; and the two worlds interact in such a
way that events in one explain ordinary events in the other. The
Nigerian tribesman is trying to do just the same kind of thing as the

Western scientist, but he starts in a different situation, and with different resources available to him. In terms of these differences we can both understand what he is doing and explain why his results differ so greatly from our own. It is this "common core of human cognitive rationality" that makes it even possible to identify the beliefs of another culture, and that in the last analysis excludes a radical relativism.

It is my view that morality ultimately depends, and should depend, on the desires and beliefs people have. If Grandy's Principle of Humanity holds good for ethics, it will claim that humans share a common core of desires and beliefs, and a common rationality in terms of which their desires and beliefs are related to one another. From this it will follow that a radical relativism is false. I would accept Lukes' arguments about the status of the Principle. That there is a common core of perception and rationality is an *a priori* truth; but the content of the common core is something that can be discovered only by empirical inquiry[7]. The importance of the Principle of Humanity, then, is that it calls to our attention the kind of effort and research needed in order to identify the moral beliefs, and hence to understand the actions, of others. In establishing that relativism is false, it underlines the complexity of the issues raised by relativists, even if their formulation of those issues was less than accurate.

I conclude, then, that discussions of a radical moral pluralism will be misleading and insecurely based unless discussants pay careful heed to the actual professional practice of anthropologists, sociologists, and scientists, rather than on popular (or philosophical) caricatures of their work. It is at least my impression that the best work in these disciplines makes it difficult if not impossible to deny that we have successfully understood a great deal about the beliefs and desires and patterns of thought of cultures very different from our own. In particular, we can understand the moral arguments and practices of people whose morality differs from ours; and the extent to which we can do this will reveal the "common core of rationality" and the common set of human desires that gives empirical content to the Principle of Humanity. The door is thereby closed on a relativist defense of a radical pluralism in ethics.

Because we are far from being able to articulate even the pattern of needs and desires that we ourselves have, and which underpin our own views about the explanation of our world, all that we can derive from this is a program with a well-founded hope of *ultimate* success. To take some examples, what *precisely* counts as an adequate moral theory, or an acceptable moral code, even in our own eyes? What *precisely* counts as an adequate theological understanding of the world? At what point do we no longer require that physics should seek yet more ultimate explanations of its fundamental postulates? (Do we need to ask why it is that nothing can travel faster than the speed of light?) Moreover, not merely can we not fully articulate which pattern of needs governs our views of adequacy in these disciplines, we are equally far from having a coherent view of why such a pattern of beliefs and needs is justifiable (insofar as it is justifiable).

It is time for some stock-taking. I have argued that relativism in ethics is false, and hence the kind of moral pluralism which relativism brings in its train. Instead, we are left with a picture of widely different, albeit ultimately related, moralities based on the widely different but ultimately related patterns of beliefs and desires that members of different cultures have. What has not been spelt out is how the fact that these different moralities are related (via the common core of desires and rationality in human nature) provides any grounds for establishing which elements in any of these moralities might lay claim to *truth*. The answer to the question "Which alterations in circumstances should make a difference to our moral behavior" in turn depends on this central question of truth.

I cannot here enter into a full discussion of the manifold problems of truth-theory. I shall simply put my cards on the table and state the basic assumptions with which I am working. I would distinguish fairly sharply between what is meant by "true" and the tests that one might use in determining truth. I take it that the best account of what is meant by saying that a statement is true is that it corresponds to the facts. But I do not believe that correspondence with the facts furnishes a workable test for which statements are true. Rather, the test for truth is a combination of coherence and pragmatism. Acceptable ways of living morally, acceptable moral codes, and acceptable meta-ethical theories will be acceptable to the extent that they are coherent and succeed in the purposes for which they are developed. I take it that (roughly, at least) a moral code consists of the set of reasons agents might give in explanation of their moral actions, and a meta-ethical theory consists in the analysis and defense of any moral code. Clearly, agents with very different patterns of life can all subscribe to one and the same moral code, since the circumstances in which they try to live by that code may well be very different in ways for which the code allows. It is less clear whether all acceptable moral codes can be accounted for by just one acceptable meta-ethical theory, though I am inclined to think that they can. In the light of these considerations, I suggest that the crucial thesis about defensible pluralism in ethics runs as follows:

(P)   There will be as many acceptable moral codes as there are different coherent patterns of true beliefs, and hence of desires that can be successfully satisfied, given the nature of the world as it is.

The Principle of Humanity guarantees that all such acceptable moral codes will be related to one another, since all will ultimately depend on the common core of rationality and desires constituting human nature. But it does not of itself guarantee that there is only one such code. The practice of anthropology indeed strongly suggests the need for caution here. We will not be in a position to judge, indeed we will not even understand, the beliefs and practices of another culture until we have succeeded in explaining how they come to hold these beliefs, given their needs and the investigative resources at their disposal. Nor will we be able to understand whether, or how far, their patterns of behavior, including their moral practices, satisfy the conditions for coherence and pragmatic success until we have understood the ways in which they have built upon the common core of rationality and desires they share with us. There is every reason to suspect that they will have built upon the common core in ways different from ours. Until the precise relationship between their ways and ours is understood, it will not be possible to understand let alone pronounce on the adequacy of their morality. If the relativist position is false, as I have argued that it is, at least it seems to me that they have achieved one major objective for which they were fighting, which was to block premature claims of cultural or moral superiority, based on the equally premature assumption that we had already reached an understanding of the needs, beliefs, and hence the moral behavior of others. If relativism is a failure, it is only a qualified failure. That relativism is false guarantees that the principles of different acceptable moral codes cannot contradict one another; but that is a far cry from saying that it is *easy* to translate beliefs formulated within one moral code into the terms of some other in order to see whether the question of contradiction even arises.

Though I have not directly argued in this paper for a realist position in ethics, it seems to me that just as relativism ultimately rests on anti-realist assumptions, so the rejection of relativism must in the end entail some version of ethical realism. The principal element in the realist position which I would endorse is that even our best view of the truth might nevertheless turn out to be defective – in contrast to the idealist view that reality is to be identified with what is asserted in our best attempts to express our beliefs. What, then, is ethical realism about? It must, I think, be a realism about human nature, about the environments in which human beings try to live their moral lives, and about the interactions between the environments and human nature which is the very stuff of our moral decisions. The various classical theories of natural law morality, from Aristotle through Aquinas, Hobbes, Butler, and Hume to such divergent authors as Kant and Mill, have all accepted the relevance to morality of each of these areas of inquiry. Plainly, our knowledge, from the realist standpoint, is notably defective. The realist diagnosis, then, of our moral disagreements does not regard the fact of disagreement, even currently insoluble disagreement, as evidence that there is no absolute truth of the matter; it is simply evidence that the current state of our moral understanding is incomplete.

2. The Variety of Human Nature

My final point is in some ways more speculative, but it is the basis for what I take to be the most radical version of pluralism which I wish to defend. Let me outline what I have in mind first, and leave some of the details until later. I suggested earlier a principle that was the basis of pluralism in ethics:

(P)   There will be as many acceptable moral codes as there are different coherent patterns of true beliefs, and hence of desires that can be successfully satisfied, given the nature of the world as it is.

I have already given some reasons taken from anthropology for believing that there will be several acceptable moral codes, and hence that (P) would justify holding a pluralism in ethics at the level of moral codes. How far this pluralism would extend depends, of course, on a detailed examination of what will count as coherent patterns, and as successful satisfaction. Such an inquiry is beyond the scope of this paper. I wish simply to explore somewhat further some reasons in support of the view that there will be in fact a plurality of moral codes to satisfy (P). The implications of (P) depend on truths about human nature, and the ways in which it relates to the various environments in which human beings live and can flourish. One aspect of human nature is of particular relevance to our topic.

Aristotle remarks that many human dispositions, those which he terms *hexeis*, are open-ended ([1], ch. 2 and 6). There is a very wide variety in the ways in which they can be actualized in human behavior. I take this observation to give us a clue to one of the most distinctive features of human nature. Precisely because of our complexity, and the rationality accompanying that complexity, there are very many ways in which we can function, and, I suggest, manifold ways in which human beings can flourish and develop their full potential as persons. Although the moral relativist in effect denies that there is any one such thing as human nature, in contrast to the moral realist who asserts that there is, it is still open to the moral realist to hold that human nature is just the kind of thing that can be variously fulfilled. My final contention is that such is indeed the case.

Let me be more exact. I am not simply saying that human beings can flourish in different environments, though they will have to behave differently in order to do so, for that says no more than that circumstances alter cases. We need different diets in different climates, different relaxations when under different pressures, and so on. I am arguing that, even in a given environment, both social groups and human individuals have a choice of structures and ways of life that can genuinely fulfill them. Within limits, we can choose which values to make pre-eminent in our lives, which pattern of social and family relationships we shall adopt, which political systems to develop. Human beings must indeed share some basic needs, and their moral and social systems must exist in part to serve those needs and to arbitrate between them when they conflict in practice. Were this not the case, we should not be able to recognize one another as human, nor would we be able to recognize as moralities the moral systems of others. The Principle of Humanity guarantees as much. But there is not, in my view, any one preferred way in which we can learn to satisfy those basic needs in a human life or in a human social structure. The reason for this, I think, is comparatively simple. It is a consequence of our rationality that we are able to think of ourselves and our world in a variety of ways. We can relate intellectually to our world from many different points of view – as physicists, as artists, as mystics, as doctors, as interior decorators. Similarly, we are able to conceptualize our most basic human urges in terms of very many different kinds of desire-patterns, according to the ways in which we order our experience generally. There will be correspondingly many ways in which humans can order their moral worlds in ways that are coherent and enable them to flourish in their environments, both as individuals and as communities.

If this is correct, then some monist views of morality will be ruled out of court at once. For example, there will be no *one* standard by which one might judge the quality of someone's life. There will be no one set of principles governing marriage and family relationships. There will be no one adequate code of justice, no one set of human rights. To be sure, there is a common humanity, and so a common underlying structure to any acceptable moral code, and also to any moral theory. There will be several very general standards of criticism that can be interculturally and interpersonally applied to moralities. That much the Principle of Humanity guarantees. But the application of these standards of criticism to particular cases will call for the greatest care. Understanding another culture, indeed understanding the life of another human being even within our own culture, is no simple matter. It is a task requiring painstaking empirical study in order to discover just how a particular pattern of behavior functions in the life of a culture or an individual, just how it is related to the desires and beliefs that are held by that culture or that individual, just how it interacts with the rest of the way of life into which it is integrated. Only when this task of understanding is undertaken will there be any secure basis for comparative moral evaluation. And even when comparative evaluation is possible, the result will be that people living in one moral framework will be able to assimilate and learn from the experience of those living in a different culture: but the result will not be one uniform morality.

To be sure, there will be negative constraints on what can claim to be an adequate morality. Individuals can be damaged in various ways, not simply by their own actions and behavior within a culture, but by the very limitations that culture places on the actions available to them. In calling attention to the common core of rationality and desires shared by all humans, the Principle of Humanity both excludes a radical relativism and also furnishes a basic set of requirements which any adequate morality must satisfy. It thereby provides a basic criterion by which any morality can be assessed, and, it may be, shown to be partially or seriously deficient. As I have already said, the tests here involve coherence and pragmatic success. The pragmatic test would focus on such things as the physical and mental health of individuals, their sense of achievement, of personal worth, and of social acceptance.

It is true that the *physical* constraints on successful human flourishing are tighter than those placing limits on our *choice* of life styles, aims, and social and political institutions. To the extent that medical ethics depends so largely on the physical features of human nature, I would expect the implications of this kind of pluralism to be less in medical ethics than in other areas of morality. Still, to the extent that some medical moral problems do depend on views about justice (as in social medicine, for instance) or about human rights (for instance, confidentiality, or the rights of parents or relatives of the patient to be involved in medical decisions), to that extent I see no reason to suppose that there is only one correct answer, or only one correct framework within which answers should be sought.

CONCLUSIONS

I have argued that ethics should be many rather than one, in the sense that several different kinds of consideration lead to the conclusion that more than one moral view can be reasonably defended. Three such considerations seem to me to be particularly important.

(1) Our lack of knowledge of the non-moral information on which our moral decisions should turn will often mean that several incompatible moral views can be reasonably defended, without our being a position to choose between them.

(2) Differences in philosophical theories about the human person, and about the relevance of non-moral information, will also mean that even when there is agreement about the non-moral facts of the case, different moral conclusions can reasonably be based on those facts. On the other hand, I have given some reasons for supposing that at least some of these differences might be lessened as the result of some recent trends in moral philosophy and in philosophy generally.

(3) Even supposing that there were complete factual knowledge, and complete coincidence of basic philosophical presuppositons about the human person and the nature of morality, I have argued that there is good reason to suppose that human nature is such as to be capable of a fairly broad range of equally worthwhile ways of ordering our individual

lives, and the structures and assumptions behind our lives together. The implication is that there is no one ideal hierarchy of values, no one ideal list of virtues, or human rights, and hence no one ideal set of moral principles. To what extent different sets of true moral principles, genuine virtues, and essential rights will overlap is a matter of empirical investigation. For it is by empirical discovery that we determine what a human being is. This is not something that can be determined *a priori*.

On the other hand, there is a crucially important sense in which I believe that Christian ethics must be one. The reason for this is that

(4) The incoherence of relativism in ethics implies that the range of acceptable moralities cannot be unlimited; and that the Principle of Humanity not merely serves to diagnose where it is that moral relativism fails, but gives the outlines of a single approach to the empirical and philosophical inquiries in which I take moral philosophy to consist.

I would like to end with a brief theological tailpiece. Views of the nature of theology that are relativist[8] fail, I believe, for essentially the same reasons for which I believe relativism generally to fail.[3] That is not to say that there is only one set of true theological statements, even ideally; but it is to deny that different theologies can be both acceptable and incommensurably different. I take this also to apply to any general theological view that lays stress on the importance of the natural law tradition in Christian ethics. There remains the challenge of accepting into our moral theology the implications of our increasing medical knowledge, our growing technological expertise, and our changing understanding of human nature in its environment and its many cultural settings. The challenge is no less than the challenge to other branches of theology posed by our growing knowledge of the historical settings of the biblical writings, or by a changed understanding of the physical universe. Our previous moral beliefs, whether in ethics or in moral theology, can in principle be falsified in the light of subsequent understanding and broader experience. A sound theology, far from being a brake on moral pluralism, must itself, and for exactly the same reasons, be many, though not incommensurably or unrelatedly many.

Heythrop College, University of London
London, England

NOTES

[1] If one wishes to speak of actions being right/wrong in themselves, I suggest that this can most coherently be done by making clear that one is speaking of *types* of action. An action will be right/wrong in itself just if any piece of behavior that is an instance of that type of action and of no other morally significant type of action is thereby right/wrong. On this account, that an action is right or wrong in itself will not of course entail that it is absolutely right or wrong. To establish that a type of action is absolutely right or wrong would require showing that any piece of behavior that is an instance of that type would thereby be wrong, no matter what other types of action it also instantiated.

[2] Davidson's original suggestion is to be found in his 'On the Very Idea of a Conceptual Scheme', originally in the *Proceedings of the American Philosophical Association* 47 (1973–4), pp. 5–20, and reprinted in J. Meiland and M. Krausz (eds.): 1982, *Relativism: Cognitive and Moral*, Notre Dame, Notre Dame, Indiana.

[3] Nicholas Lash, 'Theologies at the Service of a Common Tradition' in his *Theology on the Way to Emmaus*, (1982, Darton, Longman and Todd, London), pp. 18–33, seems to me to take the same kind of line as I do, but to concede too much to the way in which Rahner formulates his opposition to "classicism" in theology.

BIBLIOGRAPHY

1. Aristotle: 1955, *Metaphysics*, trans. W. D. Ross, Clarendon Press, Oxford.
2. Duff, R. A.: 1982, 'Intention, Responsibility and Double Effect', *Philosophical Quarterly* **32**, 1–16.
3. Grandy, R.: 1973, 'Reference, Meaning and Belief', *Journal of Philosophy* **70**, 439–452.
4. Hallett, G., S. J.: 1983, *Christian Moral Reasoning*, Notre Dame, Notre Dame, Indiana.
5. Horton, R.: 1982, 'Tradition and Modernity Revisited', in M. Hollis and S. Lukes (eds.), *Rationality and Relativism*, Basil Blackwell, Oxford, pp. 201–260.
6. Hughes, G. J., S. J.: 1980, *Authority in Morals*, Search Press, London.
7. Lukes, S.: 1982, 'Relativism in its Place', in M. Hollis and S. Lukes (eds.), *Rationality and Relativism*, Basil Blackwell, Oxford, pp. 261–305.
8. Rahner, K.: 1974, 'Pluralism in Theology and the Unity of the Creed in the Church' and 'Reflections on the Problems Involved in Devising a Short Formula of the Faith', in K. Rahner, *Theological Investigations, XI*, Darton, Longman and Todd, London, pp. 11–23 and pp. 230–246.

WILLIAM E. MAY

CAN ETHICS BE CONTRADICTORY?: A RESPONSE TO GERARD J. HUGHES, S. J.

My comments on Father Hughes' paper will focus on the following matters: (1) his crucial thesis regarding defensible pluralism in ethics, (2) the silence of this thesis with respect to the critically important issue of contradictory opposition between ethical theories, (3) his suggestion that one gather facts where there is an argument over their moral relevance, and (4) his handling of the question of personhood.

I. HIS CRUCIAL THESIS REGARDING DEFENSIBLE PLURALISM IN ETHICS

According to Hughes, "there will be as many legitimate adequate moral theories as there are coherent patterns of human beliefs and desires which can be successfully satisfied given the nature of the world as it is" ([1], p. 190). This "crucial thesis" is, in my opinion, quite impoverished as a conception of moral theory and excessively ambiguous in meaning.

It is impoverished as a conception of moral theory because "moral theories" and "coherent patterns of behavior" do not stand in a one-to-one relationship. One and the same moral theory, for instance, that of St. Thomas Aquinas, can, and does, admit that within the limits set by the moral norms grounded in that theory there can be an indefinite range of patterns of desires and behavior that can or might be successfully satisfied. In other words, while it is true that one can organize a morally good life in many ways, it does not follow that there are many legitimate and adequate moral theories and that there is no one ideal set of moral principles and norms that ought to be integrated into a legitimate and adequate moral theory. The successful satisfaction of a coherent pattern of human desires (I cannot quite see how "beliefs" can be satisfied) is not, to put the matter somewhat differently, a sign that the moral theory justifying those desires and their satisfaction is "legitimate and adequate". Within a Kantian or utilitarian or Marxist or Christian (and more of this later) ethical theory and the framework such theories provide, human beings can organize their desires in an indefi-

197

Edmund D. Pellegrino et al. (eds.), Catholic Perspectives on Medical Morals, pp. 197–201.
© *1989 Kluwer Academic Publishers. Printed in the Netherlands.*

nite number of ways and succeed in satisfying them, but all this tells us
nothing about the legitimacy or adequacy of the divergent moral theories
in terms of which the desires are organized into coherent patterns and
successfully satisfied.

This "crucial thesis" is also excessively and dangerously ambiguous.
Hughes gives no clear meaning to the key term, "satisfied". He does
refer to pragmatic success, but what exactly does this mean? If "satisfaction" just means a state or experience, then it seems to me that the door
is open to coherent ways of life that include slavery, the elimination of
the weak and nonproductive, and so forth. If Hughes wishes to include,
within adequate and legitimate moral theories, some strict limits for the
satisfaction of desires, then a substantial amount of moral theory, to be
assessed by criteria other than those of coherence and pragmatic success, will have to be brought in and the criteria for assessing them
critically examined if a determination is to be made about the legitimacy
and adequacy of the moral theory in question. This leads me to my next
point, which has to do with the applicability of the principle of noncon-
tradiction to moral theories.

II. ETHICAL THEORIES AND CONTRADICTORY OPPOSITION

All Hughes' talk about various moral theories and legitimate pluralism
within moral philosophy and, as his "theological tailpiece" adds, moral
theology, curiously evades coming to terms with the problem posed by
contradictory opposition between ethical theories. When theory X
affirms p and theory Y affirms not-p, can both be true? For instance,
some ethical theories, including, I must note, the type of natural law
theory found in authoritative teachings of the Catholic Church, hold
that some very specific moral norms are unexceptionable or, to put the
matter another way, that certain specific moral propositions are universally true, for instance, the following: "it is never morally right and good
to choose deliberately to kill newborn babies." Other ethical theories,
which might well meet the ambiguous conditions set forth in Hughes'
"crucial thesis", affirm the contradictory, namely, that "it is sometimes
morally right and good to choose deliberately to kill newborn babies".
Here we have an instance of a contradictory kind of opposition, similar
to the opposition between the following propositions: (1) "all canaries
are birds", and (2) "some canaries are not birds." The interesting thing

about propositions that are in contradictory opposition to each other is that one of them must be true and the other must be false. Both cannot be true, nor can both be false. Some specific moral norms, such as those noted above, stand in contradictory opposition to one another, and, if so, one must be true and the other must be false.

But such specific moral norms are derived from the moral principles articulated in the divergent moral theories, and the opposition found in the specific norms is traceable to a similar kind of contradictory opposition at the level of moral principles, at the core of the "moral theories". If such is the case, does it not follow that one of the theories must be true and the ones in contradictory opposition to it false? Surely Hughes must be willing to admit that the principle of non-contradiction extends to moral philosophy (and theology). If so, he must either regard the moral theory (or theories) from which the normative proposal that "it is always wrong to choose deliberately to kill newborn babies" as true and those in contradictory opposition to it as false, or else to regard this theory (or theories) false and its contradictories true.

III. GATHERING FACTS WHERE THEIR MORAL RELEVANCE IS QUESTIONABLE

Early in his paper Hughes suggests that one should proceed to gather facts when there is argument about their moral relevance. He suggests "that even where the relevance of the alleged facts of the matter is in dispute, it would still be of enormous help first to try to be clear about just what the facts of the matter really are" ([3], p. 175). He then illustrates this by saying, "Thus, prior to the dispute about whether couples should be enabled to have children independently of sexual intercourse, we ought to try to discover to what extent if any the relationships between the couple themselves and between them and their children are affected by the fact that children are conceived by different methods" ([3], p. 175). It seems to me that the same kind of reasoning would justify a couple resorting to artificial insemination by donor or to sex therapy involving genital relations with others. In short, Hughes seems to be saying that one may choose to experiment with kinds of acts one previously considered wrong, perhaps absolutely wrong. In making this suggestion he is, I submit, begging the question in favor of some sort of consequentialism or proportionalism. He is also assuming that experience of the results can verify or falsify moral norms. While it is clear that experience can verify or falsify factual statements and technical norms, it is not at all clear that experience of results can verify or falsify any moral norm whatsoever.

IV. HUGHES' HANDLING OF THE PERSONHOOD QUESTION

It is certainly true that differing concepts, both philosophical and theological, abound regarding the meaning of "person". Yet it seems to me that considerations of justice and of fair play are quite relevant in exposing the arbitrariness of some of these concepts. Many understandings of person consider it to be a quality or set of qualities enjoyed by individuals, e.g., consciousness, awareness of self as self and of others. Yet if personhood is a quality that may or may not be present in individual members of the same species, it is likewise a quality that varies enormously in degree in those members of the same (or divergent) species possessing the quality in question, so that those individuals who have more of the quality are "better" persons that those who do not. Such conceptions of persons are obviously relevant to the treatment given to individuals, whether they possess the quality or not, and it is not difficult, in my opinion, to see where this kind of thinking about persons can lead. Certainly, some reasonable assumptions about treatment of individuals, in particular, individuals of the human species, are by no means irrelevant to criticizing many conceptions of "person".

But in addition, insofar as the concept of "person" is a philosophical and theological one, and insofar as civil governments have no competence in judging the truth value of divergent philosophical and theological conceptions, the criterion, in my opinion and in that of many others, concerning the subjects who bear rights that are to be respected by societies and their members is that of membership in the human species – a question that concerns the identity of an organism within a determinate biological species. This identity is not a quality admitting of differences in degrees, for a given organism is either a member of the human species or it is not, and the determination whether it is can be made by an assessment of relevant facts with no need to adjudicate among divergent philosophical/theological concepts of "person". At any rate, I think that in his handling of the personhood question Hughes does not adequately attend to some of the substantive work that has been done in this area by many who have sought to falsify many theories of "person" by the use of reasonable assumptions appealing to principles of justice and fair play.

The Catholic University of America
Washington, D.C., U.S.A.

BIBLIOGRAPHY

1. Hughes, G. J.: 1988, 'Is Ethics One or Many?', in this volume, pp. 173–196.

PART V

PLURALISM IN SOCIETY

J. BRYAN HEHIR

RELIGIOUS PLURALISM AND SOCIAL POLICY:
THE CASE OF HEALTH CARE

The assigned topic of this paper is an analysis of religious pluralism as
the context for medical ethics and the health care ministry. In addressing this theme, I will use the situation in the United States as a case
study. There are clear limitations to this method since it will not include
relevant characteristics of pluralism which can be found in other countries and cultures. In defense of the method, my hope is that by
specifying my analysis on a situation I know in some detail it will be
possible to provide a more thorough treatment of certain aspects of
pluralism than I could if I attempted a transcultural assessment.

The argument of the paper will proceed in three steps: (1) an analysis
of the character of religious pluralism; (2) an assessment of the challenge pluralism poses for theology and ethics; (3) a specification of this
challenge in terms of health care in the United States.

I. THE CHARACTER OF PLURALISM: A SKETCH

The analysis of pluralism in the United States as a context for theology
and ethics has at least one distinguishing feature: pluralism has been the
accepted political and cultural reality in the nation since the eighteenth
century. Pluralism did not erupt suddenly in the culture as a fruit of the
Reformation, nor was it the imposed product of religious warfare.
While religious intolerance and discrimination have been present since
the earliest days of the colonial experience, the foundational constitutional documents of the late eighteenth century accepted the fact of
pluralism and shaped a polity designed to protect the religious liberty of
each person. The polity that emerged can best be described in two steps:
the structural character of pluralism and the socio-political challenge
posed by a pluralist society.

A. The Structure: A Secular State in a Pluralist Culture

Pluralism is the product of both the constitutional arrangement of
church and state as well as the socio-religious character of society in the
United States.

205

Edmund D. Pellegrino et al. (eds.), Catholic Perspectives on Medical Morals, pp. 205–221.
© *1989 Kluwer Academic Publishers. Printed in the Netherlands.*

The church-state relationship is governed by the First Amendment to the U. S. Constitution; although the word "separation" never appears in the text, the essential meaning of the Amendment is that it upholds "the separation of church and state". The combination of the "no establishment of religion" clause and the "free exercise of religion" clause in the First Amendment means that religious institutions should expect neither favoritism nor discrimination on the part of the state. The Amendment is designed to regulate the *institutional* relationship of church and state. This is a crucial aspect of religious life, but it is also a very narrow dimension of the role of religion in American society. The larger more significant question is the public role of religion not vis-à-vis the state but in society.

No idea is more centrally established in the Western political tradition than the distinction between state and society. The state is only part of society; it is constitutionally limited to specific functions and is never identified with society as such. Such an identification is the essence of a totalitarian state. However, to affirm the separation of church and state does not imply any acceptance of the separation of the church from society. It is precisely in the wider sphere of civil society that the religious communities exercise a public role through preaching, teaching, establishing educational and social institutions, and bringing their voice to bear on public policy.

The role played by the churches in the wider civil society is that of voluntary associations. While the phrase is not a theological term and it does not capture the inner meaning of a religious institution, it is a very accurate description of the public role of religious bodies in the United States. Like other voluntary agencies – labor unions, cultural organizations, professional societies – religious institutions are social groups organized for public purposes, and they are expected to influence and contribute to the public life of the nation.

There is no indication in history, law, or policy that the First Amendment was meant to silence the voice of organized religion. Although the First Amendment assures the secular character of the state, a secular state is not synonymous with a secularist society that would seek to exclude, in principle, religious insight, values, and activity in public life. A constitutionally limited secular state leaves religious institutions free to engage the debate about the content of public policy and practice in society.

The pluralist character of American society is shaped by the multiplicity of religious communities that – along with secularist philosophies – make up the civil polity of the nation. Fr. John Courtney Murray, S. J., provided a description of religious pluralism in his book, *We Hold These Truths*. In his view religious pluralism is:

> . . . the coexistence within the one political community of groups who hold divergent and incompatible views with regard to religious questions – those ultimate questions that concern the nature and destiny of man within a universe that stands under the reign of God. Pluralism therefore implies disagreement and dissension within the community. But it also implies a community within which there must be agreement and consensus ([5], p. x).

Murray's description of pluralism in 1960 included four major parties: Protestant, Jewish, Catholic, and Secularist. While these four groups still form the core of American religious pluralism, two changes have occurred since Murray wrote. First, the ecumenical movement spawned by Vatican II and cultivated in the post-conciliar period has provided a different basis for Catholic-Protestant relationships than the rather rigid categories that defined them in the 1940's and 1950's. Ecumenism has not eliminated pluralism, but it has provided motives and methods for addressing "ultimate questions" on which Catholics and Protestants still differ. Curiously enough, the ecumenical movement has enhanced both Catholic-Protestant practical cooperation on issues of social morality and Catholic-Protestant doctrinal discussion. The area yielding slowly to cooperation and common vision has concerned issues of medical and sexual ethics in their personal and public dimensions. Catholic-Protestant cooperation on civil rights and our great differences over abortion exemplify the paradox.

Second, both Murray's definition of Catholic-Protestant relations and my description of ecumenism referred to "mainline" Protestant churches (Lutheran, Presbyterian, Anglican, etc.). A basic change in the pluralist pattern has been the emergence of "the Religious Right", a theologically fundamentalist and politically conservative constituency which has as many differences with main-line Protestantism as it does with the Catholic church.[1] The emergence of the Religious Right has more political than theological significance. Traditionally, this branch of Protestantism eschewed the political arena and public activity generally as being a distraction from the Christian vocation. In the 1970's and 1980's these same denominations moved directly into political activity with the same undifferentiated analysis that led them to reject it in a former day. While their opposition to abortion gives an appearance of links to Catholicism, in fact the theological, philosophical, and ethical differences are enormous.

The pluralist character of American society in the 1980's is more complex than Murray experienced it in the 1960's. On some fronts there are deeper and stronger bonds theologically and socially; in other ways new cleavages exist, particularly on issues of public policy. The secular state, by constitutional mandate, must provide space for each of the groups to cultivate and communicate its vision and version of the right and the good for its own membership and for society as a whole. It is not the function of the state to assess the intrinsic merits of any of the competing religious claims, but to guarantee the freedom for each group to argue its cases. The influence of any group will be proportionate to its ability to persuade its own membership and others in the society of its positions.

B. The Challenge of Religious Pluralism: Shaping a Public Ethic

It is not sufficient to describe the structural character of a secular, pluralist democracy. For the purposes of this volume it is the moral challenge inherent in pluralism that is of interest. The final sentences of Murray's description of pluralism contain the kernel of the challenge:

Pluralism, therefore, implies disagreement and dissension within the community. But it also implies a community within which there must be agreement and consensus ([5], p. x).

The consensus must be created; it must be continually shaped and reshaped in light of questions facing the pluralist society. The consensus presupposes disagreement on ultimate questions but seeks limited convergence on issues of daily social significance. The substantive agenda for consensus does not touch all questions but only those bearing on the common public life of the society. It is the essence of a pluralist and a democratic society not to attempt to regulate or govern all areas of life. There must be constitutionally protected space in which individuals and groups can live out their commitment to an ultimate vision.

But some questions confront the society as a whole; it is not sufficient politically or morally to leave everything to private choice or even to separate sectarian views of what is right and good.[2] Issues touching on the common defense and the general welfare of society; issues touching on the content of civil law, the scope of state power, the protection of the weak, the infirm and the defenseless require a public response. The response will take shape in law or policy, but laws and policies in turn require some moral foundation and direction.

The consensus called for by Murray is a *moral* consensus and a *public* moral consensus directed toward issues touching the society as a whole. Not all questions in a society require a common moral perspective, and the determination about which issues require a public response is itself a crucial choice in a pluralist polity.

Shaping a moral consensus to undergird public policy involved, in Murray's view, two procedural principles. The consensus should be broad enough to provide substantive direction for public life, but it cannot be so broad that it drives any major group out of society. Determining the scope and content of the moral consensus is a never-ending task. The temptation for the pluralist society is to purchase peace among competing moral visions at the price of moral incoherence in public policy. The danger is to define the need for moral direction so narrowly that key questions remain unaddressed.

While the temptation to a moral laissez-faire attitude is a permanent one, it is difficult to sustain such a posture in the 1980's. In the twenty years since Murray wrote *We Hold These Truths*, the moral factor has become more prevalent in the public policy arena. This is not necessarily due to an increase in ethical sensitivity or moral perceptiveness; it is a product of the kind of issues on the public agenda. The explosion in medical technology that has revolutionized our capability to shape the beginning of life and to prolong the end of it; the transformation of military strategy by nuclear weaponry and changes in economic and social policy induced by global debt and national deficits, all pose questions that are not susceptible to purely technical responses. Increasingly, it is understood that on a crucial range of questions we cannot make "good policy" if we lack a conception of "right policy".

In an age in which we can do almost anything, the relevant policy question is what ought we to do. The response to this question must have a public character, it must find support in the several sectors of a pluralist society. The moral challenge of pluralism is the search for a consensus broad enough to provide moral coherence, but narrow enough to protect religious liberty.

II. THE CHALLENGE OF PLURALISM: A THEOLOGICAL-ETHICAL COMMENTARY

In one of his many commentaries on Vatican II's treatment of church-state relations, John Courtney Murray argued that *Dignitatis Humanae* and *Gaudium et Spes* jointly signaled the church's acceptance of religious pluralism as the expected context for the church's ministry ([6], p. 585 ff.). The significance of this position taken by the Council is that it marked the end of an era that began at the Reformation and continued through the democratic revolutions of the eighteenth century and the conflict between the church and liberalism in the nineteenth century. In this four-hundred-year period religious pluralism was acknowledged as a *fact* in Catholic teaching, but never accepted *in principle*. This distinction held through the pontificate of Pius XII.

The effect of this distinction was that Catholic teaching "tolerated" the fact of religious pluralism, but expended its doctrinal, pastoral, and political energy to diminishing the impact of pluralism on Catholics and to insulating the Catholic population as much as possible from engagement in a pluralist context. Even in the United States where religious pluralism and minority status were the only reality the church ever experienced, Catholic authors through the 1940's continued to reject any idea of accepting religious pluralism in principle.[3]

The shift in this posture is evident in both conciliar texts cited by Murray. The essence of the conciliar argument is found in three propositions. First, the affirmation of the right of religious liberty for every person, and the extension of the right to include freedom of worship, teaching, and witness to the communities of faith in which individuals exercise their right of religious liberty effectively dismantled the older Catholic position on religious freedom. Second, the Council limited the role of the state to a secular function; it denied to the state the right, responsibility, or competency to judge religious truth. The state's *cura religionis* is limited to protecting the right of religious liberty personally and corporately. Third, *Dignitatis Humanae* asserts that the fundamental principle governing church-state relations is the freedom of the church. The church does not ask favoritism from the state, but only the freedom to function.[4]

Taken together, these elements of the conciliar teaching do not affirm the fact of religious pluralism as good *in se*, but they do affirm pluralism as a positive possibility the church should not simply resist but should engage in the dialogical style of ministry affirmed by *Gaudium et Spes*. The post-conciliar period has involved a theological, pastoral, and political attempt to implement the conciliar view of religious pluralism.

That effort engages the church in issues of theological substance and political style. Both dimensions are relevant to the process of shaping a Catholic contribution to a pluralistic society.

The theological issue is how a religiously rooted moral vision can be articulated to influence the moral consensus of a pluralistic culture without losing its religious foundation and identity. The distinctive contribution of religious communities to the public debate lies in their religious values and moral insight. But the condition for effective moral witness in a pluralistic culture is an appeal which those who do not share a single religious conviction can find morally persuasive. Historically, Catholicism recognized this tension and relied extensively on the philosophical tradition of natural law to express much of its social teaching. The natural law ethic has been understood as a method of presenting values, principles, and policies that accord with basic insights of the Christian gospel, but do not exhaust the full range of the Gospel ethic. In explaining the place of natural law in Catholic social teaching Fr. Murray wrote:

Natural law does not pretend to do more than it can, which is to give a philosophical account of the moral experience of humanity and to lay down a charter of essential humanism. . . . It does not promise to transform society into the City of God on earth but only to prescribe, for the purposes of law and social custom, that minimum of morality which must be observed by the members of a society, if the social environment is to be human and habitable ([5], p. 297).

This traditional Catholic linkage of an evangelical ethic and its philosophical expression in natural law has itself been the subject of review and revision in the post-conciliar period. At the level of the universal episcopal magisterium there is a striking contrast in method between the 1963 encyclical of John XXIII, *Pacem In Terris* ([3], pp. 201–242), and the 1965 conciliar document, *Gaudium et Spes* ([3], pp. 243–336). The encyclical is a classic example of a Catholic natural law social ethic; there are hardly any explicitly theological arguments in the document. Two years later the principal conciliar text on social teaching, *Gaudium et Spes*, makes sparse use of natural law categories but extensive use of christological, ecclesiological, and biblical arguments in its presentation of Catholic teaching. The conciliar text on religious liberty is primarily a natural law statement of human rights and a philosophy of the state, but it includes a discreet chapter on the theological foundations of the philosophical case. A final example of post-conciliar efforts to shape Catholic social teaching for use in a pluralistic context is the two pastoral letters of the U. S. bishops on nuclear policy and the economy. Both letters seek to expose the biblical-theological basis for Catholic teaching, but in the treatment of policy recommendations for American society the letters present arguments that can be assessed in terms of rational analysis by people of differing faith and no faith ([7, 8]). The bishops distinguish in both pastorals their appeal to the community of faith (made in both theological and philosophical terms) from their appeal to the wider civil community (made in less explicitly religious terms).

In a paper on Catholic perspectives on religious pluralism, the most that can be done in 1989 is to note the various attempts being made to be more explicitly biblical and theological in presenting Catholic social teaching without forsaking the insight that a pluralistic society still needs to be addressed in the terms of a shared humanity and rationality, which are the basis of Catholic natural law thought. It is not clear as yet how the balance will be struck in this recasting of Catholic teaching addressed to civil society.

The pastoral-political challenge posed by life in a religiously pluralistic society centers on how to design a theologically coherent and politically effective position on policy issues. On areas as diverse as medical technology, military strategy, and economic policy, the dynamic of a democratic pluralistic culture calls the church to express its views on specific policies and programs.

Unless church leaders choose to remain at the relatively safe but usually vague level of stating "first principles", there quickly emerge questions of theological reasoning about how principles are to be related to policy choices. Once the church enters this domain of specifying the meaning of moral principles in concrete personal and policy choices, some classical moral questions of formal and material cooperation and toleration of evil in a social setting arise. In the last decade in the United States there have been debates among the bishops and in the wider church about policy positions on abortion legislation and nuclear deterrence which focused on these classical questions. The difficult choices arose not at the purely theological or ethical level but in the process of incarnating moral principles in policy choices.

It is clear enough that Catholic moral teaching on abortion stands in direct contradiction to existing civil law in the United States. There has been a string of explicit condemnations of the situation in the last thirteen years. A different question arises, however, when one moves from condemnation to constructive change. Here the complexity of effective action in a pluralistic context becomes evident. While a substantial percentage of American society does not agree with existing law on abortion, there is nowhere near a majority that would support a civil law incorporating the fullness of Catholic teaching on abortion.

Twenty years ago Ralph B. Potter analyzed the abortion debate in general categories that are still useful. Potter identified the "Left" of the spectrum with the position now prevalent in American law – few, if any restrictions on access to abortion. The "Right" of the spectrum Potter identified with the Catholic position – no moral justification for directly intended abortion. The "Center" he described as a position with a presumption against abortion, but allowances for specifically defined exceptions (e.g., rape, incest, life/health of the mother) ([10], pp. 85–134).

Using Potter's description it is possible to outline the pastoral-political problem faced by the Catholic position. It may be possible to build a consensus in American society today for a move from the Left to the Center, but it is not possible to construct a viable consensus to embody the position of the Right. A civil law that limited abortions to the exceptions of the Center would substantially reduce abortions in the United States. It would also allow many more abortions than Catholic moral teaching would justify. Catholic support for the position of the Center as a jurisprudential position is theologically possible, but it has attracted little support among church leaders. Such a position could "tolerate" certain exceptions in the civil law without endorsing them. Such a view could be the basis of an effective coalition, but it also could cause scandal in parts of the Catholic population. Thus far the risks of moving to the position seem more compelling to church leadership than its benefits.

An analogous problem was faced in preparing the pastoral letter on nuclear deterrence. Some argued that the elements of deterrence strategy render it intrinsically evil; others argued for some form of a "toleration" position. The pastoral letter finally invoked a judgment of "conditional acceptance" of deterrence as a "transitional strategy". The phrase was designed to criticize aspects of deterrence without condemning it. As with the abortion debate, some believe the bishops have compromised excessively with the reality of deterrence. Others see the conditional acceptance as an effective moral commentary on deterrence – the foundation of a position that many can support to move progressively toward a different posture in the nuclear age.

While the topic of pluralism within the church is the subject of another paper in this symposium, it is necessary to illustrate here how the challenge of religious pluralism is one of the sources generating ecclesial pluralism. As the Catholic community engages the wider civil society on questions of medical ethics and social ethics, the existence of both theological and political pluralism within Catholicism becomes evident.

Using the experience of the 1980's, it is possible to illustrate three levels of pluralism that have engaged the church in the United States. first, *episcopal pluralism* was clearly manifest in the 1980's debate about nuclear policy. A comparison of the French, German, and U. S. pastoral letters on the nuclear dilemma reveals a common framework of religious-moral teaching, but pluralism in *method* of moral argument and in *policy conclusions*.[5] While all three letters use the just war ethic as a starting-point, the method of argumentation varied. The French and German letters, while not identical, stressed the category of proportionality in the evaluation of deterrence. The proportional or consequentialist assessment was made both in terms of ends (the threat posed by the Soviet Union) and means (the utility of deterrence in preventing war). The American letter gave a central place to the principle of discrimination or non-combatant immunity to direct attack. This mode of argument forced the Americans to focus on the question of intention and the specific mode of targeting doctrine in U. S. strategy. The argument from proportionality played a secondary role in the American argument.

Pluralism was evident at the level of policy conclusions, particularly on the question of "no first use" of nuclear weapons, and on some of the specific conditions the U. S. letter set for deterrence.

A second level of pluralism is *theological* in the church.[6] The range of topics runs from methodological approaches to personal and policy ethics. Theological pluralism is not a new phenomenon in Catholic theology, but it has been particularly evident in the post-conciliar period. The issues run from contraception to euthanasia, from the ethics of war to the ethics of experimentation. I simply note this level of pluralism, knowing it will be treated elsewhere.

The third level is *political* pluralism within the Catholic community, not simply the obvious fact of Catholics belonging to different political parties, but also pluralism in the way Catholic medical and social ethics are applied in the public policy arena. The three major issues that have engaged the U. S. Bishops in the 1980's – defense policy, abortion, and economic policy – have all spawned articulate, sometimes organized dissent from highly visible Catholic individuals.

In 1983, Michael Novak authored an extensive article published by William Buckley in *The National Review* differing with the bishops' judgments on nuclear policy[9]. Congressman Henry Hyde, an articulate and vigorous supporter of the Catholic position on abortion, has regularly expressed his dismay and opposition to the conclusions of the war and peace letter. Senator Edward Kennedy, regularly on opposite sides on the abortion issue from the bishops, was a strong supporter of the pastoral letter.

In 1984 the pluralism of the Catholic constituency was played out for a national audience. Within the space of two months the abortion question was the subject of major addresses by Congressman Hyde and Governor Mario Cuomo. Both spoke as Catholics, both appealed to the Catholic social and moral tradition, but came down in different positions ([2, 3]).

In 1985, Mr. William Simon, former Secretary of the Treasury, led a highly visible group of laity in presenting to the U. S. public an alternative view of the ethics of economic policy from that being offered by the U. S. Bishops' pastoral letter. At the same time Gov. Cuomo and House Speaker Tip O'Neill gave addresses supporting the pastoral letter's approach.

Each level of pluralism requires more extensive treatment, but my purpose is simply to illustrate the link between religious pluralism and ecclesial pluralism. The former does not create the latter, but it provokes the debate surfacing the fact of pluralism within the church. The question posed by ecclesial pluralism is whether it erodes the witness of the church in society. In purely political terms a "united front" is always preferable, but it is also an inadequate criterion to use in assessing this question. The complexity of the issues cited in this paper and the nature of a pluralistic, democratic society guarantee that Catholic pluralism will continue.

Catholic social teaching has both expected and legitimated a certain measure of pluralism on social policy issues. The recent public attention to a range of medical ethics issues now the subject of public policy debate raises the question whether the expected pluralism on social ethics can be extended to other issues in the public arena.

The multiple issues of social and medical ethics which can spark a pluralist response within the church require systematic attention. One possibility is to think about the need for "structured pluralism" within the Catholic community. Structured pluralism begins with an affirmation that limits can be defined beyond which the adjective Catholic cannot be invoked. But structured pluralism also tends toward a modest definition of what must be commonly held on issues in the social policy debate. This view of pluralism would distinguish sharply and often between a principle and its application. It would sustain the moral prohibition of direct killing of the innocent, but know that this firm moral principle by itself yields neither a clear policy judgment on deterrence nor a single approach to the jurisprudential issues about abortion. Structured pluralism would hold firmly to a positive, active role for the state in socio-economic affairs, but would expect debate about the scope, method, and means of state intervention. Structured pluralism should be clear about the link between moral law and civil law, but it should expect differences about the amount of linkage on issues from medical questions to affirmative action legislation.

Finally, structured pluralism places requirements on Catholic teaching. The kind of explicit differentiation of levels of teaching used in the American pastoral letters points in the direction of a style of teaching which is authoritative, but not equally authoritative about all aspects of a policy question.

III. PLURALISM AND HEALTH CARE: THE SHAPE OF THE PROBLEM

The Catholic church has a multidimensional interest in the health care ministry. Without trying to review the reasons for this ministry, I will use two major areas of Catholic concern in the health care field to illustrate the kind of questions a pluralistic context poses for the church. The two areas are the classical concern of the church with issues of medical ethics and the contemporary stress within the Catholic health care community with serving the poor. Both issues, medical ethics and social justice concerns, are rooted in Catholic teaching. Both are today at the forefront of the health care debate. Each of them poses distinct questions and problems for the church's teaching and witness.

The range of medical ethics questions engaging the Catholic church today will undoubtedly be addressed by other authors in this colloquium. My interest is to highlight the fact the most of these classical moral issues (from reproductive ethics to experimentation, to care of the terminally ill) are today also public policy questions. The moral issues are increasingly also legal and political issues.

Three factors have been principally responsible for transforming the traditional medical-moral questions into public policy concerns. First is the rapidly expanding role of the government in medical research and medical care. The growth of the government's role in every area of the medical system, from sponsoring research on pathbreaking technologies to providing funding for medical care for the poor, requires that explicit public policies be debated and adopted to guide the role of the government in these profoundly human questions. Second, there has been a substantial increase in the aspects of medical care now covered by civil law. Medical treatment involving the beginning of life and the end of life, research that could violate basic human rights, procedures that are morally offensive to part of the population and accepted as good medical practice by others, have all put pressure on the legal and judicial system to become increasingly involved in health care questions. Third, it is precisely the moral divisions within American society at large – among legislators, health professionals, academics, and citizens – which have pressed the medical-moral issues to center stage in the political arena. Because individuals or groups fear future developments (e.g., in genetics) or are already in disagreement with existing practice (e.g., abortion), they seek to control and direct medical research and medical care through the civil law and public policy. The debates about the proper role of the government, the extent or limits of the civil law in medical care and moral divisions of the society take place within the conditions of pluralism outlined earlier in this paper. The search for moral consensus to guide law, policy, and practice in medical research and health care is today one of the most conflicted arenas of the pluralist society. The long interest of the Catholic church in medical ethics has today taken on new public significance as other institutions, groups, and individuals join the moral argument.

A different kind of moral question is posed by health care and the poor. Catholic social teaching, which today calls for a "preferential option" for the poor, presses the church to see the question of access to health care in social justice terms. In the United States health care has not been addressed in terms of a well-defined social policy. Other areas of human need – nutrition, unemployment, housing – are far from adequately addressed, but the attempts to deal with them have been more systematic. There has been a greater recognition that these areas of need fall within the scope of public responsibility. It is much easier to argue for the "right" of basic nutrition or even the "right" to a job than to argue for a "right" to health care in the United States.

The health needs of the old, the young, and the destitute, however, have forced themselves on the public consciousness. The response has been embodied in existing programs of Medicare and Medicaid. But a convergence of factors, some of them macroeconomic in nature, others more directly related to changes in the financing of health care, have converged in the 1980's to make the situation of the poor very precarious. A recent report of the Catholic Health Association, *No Room in the Marketplace: The Health Care of the Poor* ([1], pp. 2–3; 5–6) documents the shape of the problem:

– 34 million people are without health insurance for all or part of the year;

– for a variety of reasons, the number of persons potentially unable to pay for health insurance has risen steadily since 1979;

– budget pressures and the continuing influence of the national deficit make legislatures unwilling to think of increased expenditure for health care;

– the impact of health costs reaches beyond those defined as poor to middle class families; health care costs can be decisive in moving families into poverty.

For a church that defines health care as a ministry not a business, the problem of health care and the poor is a policy question of social justice. Hence the response to the problem, exemplified by the report *No Room in the Marketplace*, is both an examination of what Catholic health care facilities are doing directly and an analysis of how the church should enter the wider public policy debate about health care financing and health care policy as it affects the poor.

Both medical ethics and social justice aspects of health care are today public policy issues. Both issues today are debated and significantly influenced by civil law and government policy. On both questions the Catholic church enters the public debate with a complex framework of moral and social teaching. The church has an interest in both issues and a quite definite policy approach to both.

In a pluralistic society, the church, as we have seen, brings its perspective to a wider arena where a consensus – political and moral in nature – must be shaped. The interesting policy problem for Catholic teaching arises at the point where the medical ethics issues and the social justice agenda intersect.

From the perspective of several issues in medical ethics today (e.g., *in vitro* fertilization, sterilization, abortion), official Catholic teaching is either opposed to existing law and policy or seeks to restrict the effects of certain policies (e.g., protecting Catholic institutions from being forced to perform certain procedures). While the maximal objective of Catholic policy in several of these areas is to shape the wider social consensus to accord with Catholic teaching, the practical goal is often to provide "space" for Catholic institutions through conscience clauses or exempting amendments to legislation. In general terms Catholic policy, from the perspective of medical ethics, is to restrict the state's role, to guard against governmental intervention, which would create a moral crisis for Catholic institutions that are often the beneficiaries of public funding.

From the perspective of the social justice agenda, the growing demands of health care for the poor make it clear that an increased role of federal and state government is necessary if an adequate public policy is to emerge. On this front, Catholic social teaching, which has a strong sense of the moral responsibility of the state, is often inclined to advocate greater federal and state involvement in health care.

The intersection of the medical ethics and social justice issues, therefore, is principally but not exclusively focused on the role of the government in relation to private sector institutions in health care. Correlating Catholic concerns about state policy on medical-moral issues and Catholic interest in a more aggressive public role on social justice issues would be a complex task in any case. But the dynamics and demands of religious pluralism intensify both sides of the policy equation.

This paper is not the place to design a comprehensive response for Catholic institutions and Catholic moral teaching. Some comments for the direction of a coherent policy are possible. First, the basic challenge is to address both sides of the problem without sacrificing one to the other. Addressing both issues is in the first instance a conceptual or definitional question; it would be all too easy to define the "moral issues" as the medical-moral topics, while defining the social justice concerns as "political" issues. Such a definition will inevitably give policy priority to the first agenda.

Second, if both sides of the equation are kept in focus conceptually, the next step is to establish the essential objectives in each category – what cannot be compromised – and to distinguish these from desirable but negotiable objectives. It seems clear, for example, that a national health insurance policy that would meet the needs of the poor should receive strong Catholic support. It is also clear that the inclusion of abortion in such a policy would create an insurmountable obstacle for Catholic support. But what if the policy did not include abortion, but did include sterilization. It is not self-evident that it should draw Catholic opposition.

Third, some specific provisions – conscience clauses – will have to be a continuing concern in law and policy for Catholic institutions; these should also be concerned with protecting individuals from conflicts of conscience in the performance of professional duties.

Fourth, some "priority principles" need to be worked out to resolve more complex cases of the problem raised in point two, i.e., when the demands of the social justice agenda conflict with the demands of the medical-moral agenda.

The context of religious pluralism will set the conditions in which the church responds to both sides of the health care problem. It will also be a catalyst generating new questions for Catholic theology and ethics precisely because religious differences will yield different conceptions of how to address medical ethics and standards of justice. The resources to engage in the full range of the health care debate and to engage confidently in the dialogue of a pluralist society are readily available in Catholic medical and social ethics. The disciplined use of these resources is the future challenge.

Kennedy Institute of Ethics
Georgetown University
Washington, D. C., U. S. A.

NOTES

[1] For a discussion of the Religious Right from a sympathetic but critical perspective cf. R. Neuhaus: 1984, *The Naked Public Square: Religion and Democracy in America*, Eerdmans, Grand Rapids, Michigan.

[2] For arguments about the need to move beyond a purely individualistic and "minimalistic" ethic cf. D. Callahan: 1973, *The Tyranny of Survival*, Macmillan, New York and, 1981, 'Minimalist Ethics', *The Hastings Review* (October), 19–26.

[3] Cf. J. A. Ryan and F. J. Boland, C. S. C.: 1940, *Catholic Principles of Politics*,

Macmillan, New York. and J. Fenton: 1974, 'The Theology of Church and State', *Proceedings of the Catholic Theological Society of America* **2**, 15–46.

[4] The historical background for this shift of position is traced in J. C. Murray, S. J.: 1964, 'The Problem of Religious Freedom', *Theological Studies* **25**, 503–575.

[5] The French and German letters can be found in J. Schall: 1984, *Winning the Peace*, Ignatius, San Francisco.

[6] For a discussion of pluralism in Catholic theology cf: C. Curran: 1975, *Ongoing Revision in Moral Theology*, Fides/Claretian, Notre Dame, Indiana, ch. 2.

BIBLIOGRAPHY

1. Catholic Health Association: 1986, *No Room in the Marketplace: Health Care and the Poor*, Catholic Health Association, St. Louis.
2. Cuomo, M.: 1984, 'Religious Belief and Public Morality', *Origins* **14**, 234ff.
3. Gremillion, J. (ed.): 1975, *The Gospel of Truth and Justice*, Orbis Books, New York.
4. Hyde, H.: 1984, 'Religious Values and Public Life', *Origins* **14**, 266ff.
5. Murray, J. C., S. J.: 1960, *We hold These Truths: Catholic Reflections on the American Proposition*, Sheed and Ward, New York.
6. Murray, J. C., S. J.: 1966, 'The Issue of Church and State at Vatican II', *Theological Studies* **27**, 580–606.
7. National Conference of Catholic Bishops: 1983, *The Challenge of Peace: God's Promise and Our Response*, U. S. Catholic Conference, Washington, D.C.
8. National Conference of Catholic Bishops: 1986, *Catholic Social Teaching and the U. S. Economy – Third Draft, Origins* **16**, 33ff.
9. Novak, M.: 1983, *Moral Clarity in the Nuclear Age*, T. Nelson, Nashville, Tennessee.
10. Potter, R. B.: 1968, 'The Abortion Debate', in D. R. Cutler (ed.), *Updating Life and Death: Essays in Ethics and Medicine*, Beacon Press, Boston, pp. 85–134.

WALTER J. BURGHARDT

CONSENSUS, MORAL WITNESS AND HEALTH-CARE
ISSUES: A DIALOGUE WITH J. BRYAN HEHIR.

As always, Bryan Hehir's paper is packed; there is hardly a wasted
word; the argumentation is at once as rigorous and flexible as the man
himself; and there is a discouraging absence of weak spots where a
commentator might call him to logical or theological account.

Helpfully, Father Hehir has gathered his reflections under three
rubrics. First, he sketches the character of religious pluralism, "the
accepted political and cultural reality in the nation since the eighteenth
century," and focuses insightfully on "the moral challenge inherent in
pluralism." In a pluralist and democratic society, how do you create,
how do you continually shape and reshape, a public moral consensus
that will take concrete form in law or public policy acceptable to the
community – a consensus "broad enough to provide substantive direction for public life, but [not] so broad that it drives any major group out
of society"?

Father Hehir's second major point is a theological-ethical reflection
on that challenge. After noting the "new thing" on the Catholic side,
that is, Vatican II's affirmation of religious pluralism not "as good *in se*"
but "as a positive possibility which the church should not simply resist
but engage" in dialogical fashion, he asks how Catholicism can bear
effective moral witness within such a culture and appeal persuasively to
those who do not share Catholicism's religious convictions. He is aware
that historically Catholicism has "relied extensively on the philosophical
tradition of natural law to express much of its social teaching."
Twentieth-century exemplars of this approach are John Courtney Murray and John XXIII's 1963 encyclical *Pacem in terris*. More recently,
efforts have been made "to be more explicitly biblical and theological in
presenting Catholic social teaching without forsaking the insight that a
pluralistic society still needs to be addressed in the terms of a shared
humanity and rationality which are the basis of Catholic natural law
thought." Examples: the two pastoral letters of the U.S. bishops on
nuclear policy and on the economy. "It is not clear as yet," Father Hehir
observes, "how the balance will be struck in this recasting of Catholic
teaching addressed to civil society." He indicates how, in the last

Edmund D. Pellegrino et al. (eds.), Catholic Perspectives on Medical Morals, pp. 223–229.
© *1989 Kluwer Academic Publishers. Printed in the Netherlands.*

decade, difficult choices have arisen for the bishops and the wider Church, "not at the purely theological or ethical level but in the process of incarnating moral principles in policy choices." Prime examples: abortion legislation, nuclear deterrence, and economic policy.

Father Hehir's third rubric relates American pluralism to the health-care ministry. He uses two major areas of Catholic concern to illustrate the challenge of pluralism: issues of medical ethics and service of the poor. The traditional medical-moral questions have been transformed into public-policy concerns principally by three factors: debates about the proper role of government, the extent or limits of civil law in medical care, and moral divisions within American society at large. In each of these areas Catholicism is agonizingly involved in a search for moral consensus to guide law, policy, and practice in medical research and health care. The problem of the poor, Father Hehir asserts, "presses the church to see the question of access to health care in social justice terms." How should the Church "enter the wider public policy debate about health care financing and health care policy as it affects the poor"? For Hehir, "The interesting policy problem for Catholic teaching arises at the point where the medical ethics issues and the social justice agenda intersect." The intersection "is principally but not exclusively focused on the role of the government in relation to private sector institutions." Here Catholic policy tends to restrict the state's role in medical ethics, expand it in care for the poor. Father Hehir recommends: (1) address both sides of the problem, the medical-moral and the social-justice, without sacrificing one to the other; (2) distinguish essential objectives from desirable but negotiable objectives; (3) insist on provisions for the rights of conscience in law and policy; (4) work out "priority principles" to resolve more complex cases "when the demands of the social justice agenda conflict with the demands of the medical-moral agenda."

Finally, Father Hehir is convinced that in Catholic medical and social ethics we have the resources to engage fully in the health-care debate "and to engage confidently in the dialogue of a pluralist society." The challenge, from now on in, is "the disciplined use of these resources."

I do not think it unfair or uncharitable to suggest that Father Hehir has not solved the problems he has raised. Such was not his intention; such was not his achievement. He has sketched, with consummate care, the problems of principle and policy, of ethics and law, that health care faces and will face in a society discouragingly divided on issues of ultimate concern and in day-to-day living. He has put health care in its proper American perspective: religious pluralism. In so doing, he has raised our consciousness, our sights, and our hopes.

What, then, is left for a commentator, if he is not simply to fold his theological tent and steal silently away? Not properly a response, save a vigorous amen to Father Hehir's major points. Rather a brief reflection triggered by specific moments in the paper. Actually, three moments, and three reflections. Not so much in dissent; call them addenda, or calls for clarification.

First, the paramount issue of *consensus*. As Father Hehir sees it, the consensus must be created, must be continually shaped and reshaped; it presupposes disagreement on ultimate questions, seeks limited convergence on issues of daily social significance. I am puzzled on two counts: (1) I am not clear on how "ultimate" the disagreement presupposed may be, what "ultimate" means in this context; and (2) I much fear that disagreement on "ultimates" is precisely what often makes for divergence on issues of daily social significance.

Let me concretize my problem by harking back to John Courtney Murray, to whom Father Hehir is admittedly indebted but not enslaved. With the Founding Fathers, Murray held that there exists an ensemble of substantive truths that "command the structure and the courses of the political-economic system of the United States" ([3], p. 106), truths that can be known by reason – not indeed self-evident but reached by "careful inquiries" of "the wise and honest" ([3], p. 118). Reduced to its skeleton, the consensus affirmed a free people under a limited government, guided by law and ultimately resting on the sovereignty of God.

Does the consensus still exist? Not really, Murray argued. Especially if you combine the consensus with its basis in natural law. "By one cause or another it has been eroded" ([3], p. 86). Influenced by modern rationalism and philosophy, "the American university long since bade a quiet goodbye to the whole notion of an American consensus, as implying that there are truths that we hold in common, and a natural law that makes known to all of us the structure of the moral universe in such wise that all of us are bound by it in a common obedience" ([3], p. 40). For its part, Protestant theology has never been happy with the thesis of a human reason so sheltered from original sin that it can know God unaided by grace. Perhaps the people are wiser than their philosophers and pastors, but such a hope Murray found too "cheerful" for his intellectual comfort. As for Roman Catholics, traditionally their "participation in the American consensus has been full and free, unreserved and unembarrassed, because the contents of this consensus – the ethical and political principles drawn from the tradition of natural law – approve themselves to the Catholic intelligence and conscience" ([3], p. 41). Regrettably, within our philosophically and religiously pluralist society we do not have a common universe of discourse: we do not know what the other is talking about.

In consequence, my first question: What, if any, is the relationship between the consensus whose absence Murray deplores and the consensus on ultimates whose absence you find compatible with a flourishing pluralist society? For a time I thought that the answer might be: Murray's consensus is political, yours is religious. But Murray's consensus is not nakedly political; it is profoundly moral. So then, is the difference a nuance between fundamental morality (where we ought not disagree) and religious ultimates (where we may disagree)? But if so, aren't we in deep trouble? Can we have disagreement on fundamental morality and still have what you call "limited convergence on issues of daily social significance"?

Father Hehir's Response:
There are three levels of argument in a pluralist society where one can test the possibility of and limits of "consensus"; they are (1) religious, (2) moral, and (3) political (involving policies, laws, and programs).

Murray held, and I agree, that the lack of religious consensus is the defining characteristic of a pluralist society. We both accept that as a given. The challenge of the pluralist society is to probe the possibilities for moral consensus and political consensus in light of the presumed lack of religious consensus.

Murray thought moral consensus *should be possible* in American society, based on "the tradition of reason" in public affairs. He concluded, regretfully, in *We Hold These Truths* that moral consensus *did not exist*. But I have always taken the abiding challenge of Murray's work to be both a goal and an inspiration to continue the search for moral consensus. In other words, his assertion that consensus is possible is for me an imperative to pursue the possibility. His assessment that consensus is lacking is for me an empirical judgment that can be changed.

The political level of a democratic society presumes differences of view on specific policies and programs. But beneath the differences on contingent issues, there needs to be a basic direction for politics and law that provides a foundation for coherent governance. This foundation, just below the political level, is not a purely moral judgment, but it requires some moral argument. On issues as diverse as nuclear policy, abortion, and human-rights policies, failure to achieve some moral agreement can block the development of effective policy.

My view is that disagreement on "ultimate issues" is found at the religious level; existing disarray at the moral level is increasingly seen as incompatible with effective governance. In other words, we don't have moral consensus (Murray is still right) but an increasing number of people see the irreducible necessity of it. Hence my view is, moral consensus is possible, should be a priority item in public debate, and must be pursued at both the moral and political level simultaneously.

A second reflection, relative to Father Hehir's remarks on *how* the Catholic community might most effectively *influence* the moral consensus of our pluralistic culture. In 1979 he contributed to a discussion in a distinguished theological journal on "Theology and Philosophy in Public: A Symposium on John Courtney Murray's Unfinished Agenda" [2]. In that symposium, John Coleman and Robin Lovin suggested that Murray's efforts to renew the American public philosophy can and should be supplemented by a public discourse that explicitly appeals to Christian symbolism. Father Hehir called for a reappropriation of Murray's method as indispensable in today's situation, and he stressed the need for a renewed public philosophy if both America and the American Church are to move intelligently toward greater justice in a world marked by deep pluralism of ultimate beliefs. David Hollenbach concluded that neither an exclusively particularist public theology nor an exclusively universalist public philosophy will serve the needs of the Church at this historical moment. The task of fundamental political theology, he claimed, is to discover the relationship between these two spheres of meaning.

In consequence, my second question: Though your formal paper does not take a position on this issue, would you mind walking out on that particular limb? What say you now? What you said in 1979, or have you experienced a "development of doctrine"?

Father Hehir's Response:
I remain interested in the public theology proposal of David Hollenbach and John Coleman, but I remain convinced that in social ministry, at the point the Church enters the policy debate, the leading edge of our position should be shaped by Murray's notion of public philosophy. I see a twofold role for public theology: (1) shaping an internal consensus in the faith community about the main components of societal life and policy; and (2) offering the wider society a perspective of the foundational insights (e.g., the prophets, Jesus' life, ministry, and teaching) that shape the Church's approach to questions. Many in the wider society may find this perspective useful for insight or inspiration – we should *offer* it to them. But when we argue that certain policies should be adopted, certain actions prohibited, or certain laws enacted, I maintain the pluralist dialogue requires an expression of our views in the style of public philosophy.

A third reflection, in connection with Father Hehir's third rubric,
"Pluralism and *Health Care.*" I mulled over this section with especial
care, because I have been increasingly concerned through fifteen years
to fashion a theology and spirituality of health care, and because I am
chairman of the recently formed Committee on Theology and Ethics of
the Catholic Health Association. Like Father Hehir, I have been profoundly moved by the CHA report *No Room in the Marketplace: The
Health Care of the Poor*[1].

I am concerned, of course, with the broader pluralism that often
opposes Catholic doctrine to existing law and public policy. But as I
roam across our country in what a colleague called "the hoof and mouth
disease", I am even more concerned over our internal Catholic pluralism. This in both of the health-care areas that dominate Father Hehir's
third point: medical ethics and social justice. I am not surprised that
issues of medical ethics divide us – divide magisterium and theologians,
divide theologians among themselves, divide theologians and the theologically unsophisticated. Complex issues are at stake; authority clashes
with argument; life and death meet on a hospital bed; passions run high.
What dismays me is that there are still a substantial number of Catholics
who do not believe that the Church has a proper mission at all in the
social arena, who resist unto blood and money a "preferential option for
the poor", who are convinced that the Church's commission is to gather
a band of true believers who will prepare themselves by faith and hope
for the redemptive action by which God will establish the kingdom at
the end of history. Parish priests tell me that parishioners tune them out
when they preach the pastorals on peace and on the economy. Is health
care next on the "hit list"?

In consequence, my third question: From your several vantage points,
Father Hehir, do you see our internal Catholic dissension on medical
ethics and social justice as a serious obstacle to our efforts to influence
public policy, to shape the wider social consensus? And if you do, would
you write up a prescription ("three times a day for a week") that might
cure these ills and make for a healthier Catholic body?

Father Hehir's Response:
Undoubtedly there are differences within the Catholic community on both the involvement of the Church in public ministry and on specific positions taken. The two pastoral
letters distinguish these two levels. The letters expect, invite, indeed seek to promote
dialogue, debate, and assessment of specific positions advocated in the letters (and in
other documents, congressional testimonies, etc.); disagreement at this level is both

inevitable and can be conducive to sharpening the institutional Church's views on specific questions.

Fundamental disagreement on the social ministry as such is quite another problem. It does exist, but I do not find it as pervasive or profound as your question indicates. I find an openness to an activist church on public questions. I believe we have a long way to go in building an ecclesial consensus on social ministry, but I think the most striking shift in Catholic thought and practice in the public arena since Vatican II has been an acceptance of social witness as a "constitutional" dimension of Catholic life. We need to go farther, but we have already come a long way.

My personal thanks and our corporate gratitude to you, Father Hehir, for being, as always, intelligently informative and passionately provocative.

Georgetown University,
Washington, D.C., U.S.A.

BIBLIOGRAPHY

1. Catholic Health Association: 1986, *No Room in the Marketplace: The Health Care of the Poor*, CHA, St. Louis.
2. Hehir, J. B.: 1979, 'The Perennial Need for Philosophical Discourse,' in 'Theology and Philosophy in Public: A Symposium on John Courtney Murray's Unfinished Agenda', *Theological Studies* **40**, 700–715.
3. Murray, J. C., S. J.: 1960, *We Hold These Truths: Catholic Reflections on the American Proposition*, Sheed and Ward, New York.

ROBERT F. LEAVITT

NOTES ON A CATHOLIC VISION OF PLURALISM

In the framework of this conference on Catholic Perspectives on Medical Morals, the subject of pluralism and the Church seems to demand
complementary reflections *ad intra* and *ad extra*. This paper will focus
on the subject *ad extra* by inquiring about a Catholic vision of the
phenomenon of pluralism in our modern social world.

These reflections are deliberately called "notes" to avoid the impression of a unified Catholic theory of pluralism. Instead, they constitute a series of observations on the meaning of pluralism from a Catholic
perspective, drawing also on certain historical and philosophical currents of thought, which seem to the author to join forces with or
properly qualify the Catholic point of view. That point of view, it seems
to me, is marked essentially by Catholicism as a quest for unity and
order in revelation and reason. By unity I mean a productive integration
of the various "orders of truth" (Ricoeur's phrase) in human experience. Therefore, the phenomena of pluralism are inevitably evaluated
in relation to this intention. Catholicism aims to witness to the unity of
truth, in spite of what seems in our time to make this a hopeless project.
Sometimes it witnesses to it through an Augustine's vision of historical
meaning (*The City of God*) or through a Thomas' synthesis of reason
and faith (*Summa Theologiae*) or through a conciliar re-structuring of
doctrine (Trent and Vatican II) or through counter-cultural proclamation (Vatican I and recent post-Vatican II experience).

The paper will begin with some historical reflections on the meaning
of order and pluralism for a Catholic vision because the signicance of
these terms shifts with changing intellectual contexts. From this, we will
move to a consideration of the internal grounds in Catholic thought for
an affirmation and qualification of pluralism in faith and culture. Finally,
we will reflect on some problematic meanings our contemporary pluralistic situation has for Catholic thought.

I. ORDER AND PLURALISM IN AN HISTORICAL VIEW

In the preface to his work *Order and History*, Eric Voegelin writes:
"Every society is burdened with the task, under its concrete conditions,

231

Edmund D. Pellegrino et al. (eds.), Catholic Perspectives on Medical Morals, pp. 231–253.
© *1989 Kluwer Academic Publishers. Printed in the Netherlands.*

of creating an order that will endow the fact of its existence with meaning in terms of ends divine and human" ([14], p. ix). Voegelin views human history fundamentally as the attempt to create symbolizations of order under different circumstances. He sees a movement from myth to revelation to polis to philosophy to empire to Christianity to national states to gnosis. His sequence is not important for our purposes except that it shows the variety of symbolic forms that order may take: myth, bible & people, society, religion, the state, ideology. These forms, in Voegelin's theory, move from a more compact symbolism to a more differentiated symbolism of order. Symbols of order, in the course of time, may disintegrate, or what Voegelin calls a "leap in being" may occur, giving rise to a new symbolic configuration. *Pluralism, then, only makes sense within a given symbol of order or a loss of the same or as part of an emergent symbolism of it.* In this context, I would like to look at four instances of the Catholic response to pluralism in different historical symbolizations of order.

1. Diversity to Order: Early Catholicism

The final phase of the New Testament period has been described by the term "early" or "emergent" Catholicism. The phenomenon is represented in the New Testament by the Pastoral Letters, James, 1 and 2 Peter, and Jude. Outside the canon, such works as 1 Clement, Ignatius, and the Didache are examples of it. They all witness to the gradual institutionalization of Christianity as the lively expectation of Jesus' imminent return started to fade away. The sociological manifestations of early Catholicism are the establishment of an organized ministry and hierarchy, an authoritative canon of writings, and a formal creed.

With the appearance of Gnosticism, a complex religious amalgam of Christian, Jewish, and Hellenic speculations, the Christian Church found the need to articulate its creedal position more clearly. Having defined itself earlier over against the synagogue, Christianity now had to define itself in opposition to certain internal Christian movements. Thus, the appearance of what came to be called orthodoxy and heresy – concepts that are reciprocally defined – begins to develop. Historians, like Walter Bauer[1], have helped us to understand and appreciate the rather shadowy boundaries that actually existed between certain early Christian groups. Prior to the bifurcation of ideas and movements into "orthodox" and "heretical" – such as in Irenaeus' *Adversus Haereses* – we learn that concepts later declared heterodox were in fact regnant concepts in certain early Christian communities. So, the historian of religion defines Christianity in this early stage not by the standards of the later creeds or the writings of the majority party, but by the multiple voices, however incoherent or exceptional by later criteria, which considered themselves Christian.

But, in this array of early diversity, the historian also discerns the trajectory of unity, order, and authority in the Church. From the late New Testament period through the sub-Apostolic writings to Irenaeus and the early Christian apologists, there is a rising concern for discriminating truth and falsehood in purported Christian teachings. The concept of the Apostolic Tradition is developed to insure the continuity of teaching through the episcopacy in local churches. From this, truth means "what has been passed on," and falsehood is what is novel, different, unknown in tradition.

In Troeltsch's[13] distinction between church and sect,[1] Christianity later evolved from the social status of a sectarian movement to a church by virtue of its public establishment as an institution in the Roman empire. The conversion of Constantine symbolized Christianity's new public status. One of the traits of the church-type institution is its engagement with the surrounding culture and its willingness to absorb ideas from these surroundings. The gradual movement of Christian theology beyond apologetics to theological exegesis of the Bible and the use in this of secular philosophical systems signals an openness to the non-Christian thought-world. Intellectual ferment in the Church grew by bringing biblical revelation and philosophical concepts together. So, alongside the Catholic concern for the unity and identity of faith (first sharpened by the conflicts with Gnosticism), we also find an emergent ecumenism in Catholic thought and the beginnings of a philosophical hermeneutic of Christian faith. By adopting conceptual schemata from late Roman and Greek culture, the Church achieved a new language and the basic conditions for influencing the culture in return. The development of a Christian culture in the West had its roots in this period marked by the passion for Christian identity and the openness to cultural diversity.

2. Synthesis to Conflict: Middle Ages to Reformation

In its intellectual activity, the Middle Ages was constructed on the Catholic neo-Platonic worldview of Augustine and its mystical corollaries filled in by Dionysius (whom the mediaevals thought was the companion by the same name of Paul in *Acts*, and whose writings, therfore, they invested with near apostolic authority) in his *Celestrial Hierarchy* and *Ecclesiastical Hierarchy*. Beyond this legacy, mediaeval dialetics refined the procedures of theological argumentation beginning with Peter Abelard and Peter Lombard. Thomas Aquinas took both of these traditions and synthesized the *lectio divina* with the newly translated works of Aristotle to produce his theological system. He took theology beyond the stage of biblical commentaries, homilies, and occasional works to the status of a scientific (that is to say "theoretical") discipline. As with all scientific thought, theology had then settled on a paradigm for testing opinions. It established an orderly investigative method, more sophisticated than Abelard's *Sic et Non* – which juxtaposed conflicting opinions received from tradition – that amounted to a logic of theological categories. Aquinas developed this logic, using Aristotle's metaphysics and ethics, for resolving incommensurable terms in Christian faith such as grace and freedom, creator and creation. The philosophy of being and its analogical application was used to reconcile the differences between revelation and reason.

The Thomistic synthesis, of course, was only one of many theological schools in the mediaeval world. Bonaventure, Scotus, Occam all produced theologies that differed from the Thomistic system. Because of this, scholasticism is often called pluralistic in having produced rival schools of thought and intense debate. Still, we must remember that scholastic controversies occurred in a culture permeated by the Christian vision of things. The Christian mediaeval culture permitted the divisions within scholasticism, and, one might say, actually facilitated an increasing subtlety in disagreement, making the disputes as esoteric as they would later seem to be irrelevant. Scholastic pluralism is only an imperfect analogue of our own situation.

The Renaissance and Reformation, though quite continuous with the mediaeval world in many respects, signalled the break-up of the unity of empire and church, reason and faith. This happened because the cultural structures sustaining the mediaeval sense of unity and an ordered-plurality of theologies began to dissolve. In its place was born a new vision of the world partly drawn from classical sources, partly from the Bible, and partly from new scientific-geographic discoveries. The symbol of the "New World", with its rich and exotic natural cornucopia, forecast a new intellectual age. Explorers brought back chimpanzees, hitherto unknown in Europe, so humanists like Thomas More could own one – the Renaissance equivalent of the moon rock! Nature was now eclipsing revelation and speculative reason as the source of knowledge about man and universe. Sciences of nature were about to replace the science of the divine and its hierarchical universe. The full conflict had to wait until Galileo espoused the heliocentric theory (though Copernicus had developed it as early as 1510–1514) with data from his telescope and mathematical calculations. The founder of modern physics and mechanics, Galileo began to unify celestial and terrestrial phenomena is a single theory. He undertook the mathematization of the natural world – later advanced by Newton and Descartes – which set the stage for modern science and a decisive shift in the meaning of truth. The old definition of truth was still the same – *adequatio mentis ad re* – but the "reality" was shifting from that defined theologically to that emerging from experiment. The social definition of reality was changing as well as the evaluation of the intellectual tools fitted to its exploration.

From the Renaissance on, science slowly established itself independently of theology which, in time was forced to retreat from its earlier commanding intellectual position. The stage was then set for a scientific definition of reality as a whole.

Thus, as the order of science began to replace the celestial order of revelation, a new unity emerged to challenge the pluralism of the schools. Alongside this unity, a new political pluralism developed in the emergence of national states from the remains of the Holy Roman Empire.

The religious pluralism of the Protestant Reformation overlapped with this new national consciousness, and drew much of its energy from it. In some ways, Protestantism was no more radical than earlier mediaeval conflicts. The normative theology of the Church, which respected the institutionalization of the papacy and tradition, was challenged by a new theology of Scripture. The Bible became the critical counter-principle to papacy and tradition. The Protestant challenge to the Church led to the Council of Trent and the Counter-Reformation effort, which eventually consolidated a new Catholic consciousness determined largely by the polemics of the Reformation. In the same period, Protestantism developed its own versions of Lutheran and reformed scholasticisms preoccupied with the intellectual consolidation of Protestant consciousness. Hence, an ecclesial and theological pluralism within Western Christianity came into being with the Protestant Reformation.

This experience of pluralism in the Reformation looked, to the indignant eye of 17th and 18th century thought, as only a variation on mediaevalism. If the reformed churches thought of themselves as a burst of the spirit of freedom, to another more skeptical viewpoint they were only an offspring of the parent faith, just as dogmatic and intolerant as the wars of religion made clear. The rationalists appraised the Reformation in its effects: the unity of society would never again be possible on the basis of religion. A new symbol of order was in the making: autonomous reason.

3. Reason to Criticism: Enlightenment and the 19th Century

The Enlightenment set itself the task of conquering the pluralism of historical religions (with their human traditions) with a new vision of unity. The unity of autonomous reason – deductive or inductive, scientific or philosophical, practical or critical – would be the new foundation of a new culture. The unity that was impossible from religion because of its inevitable dogmatism would be possible through the independent, free exercise of reason. For its religion, the Enlightenment invented "natural religion", a rational faith that discerned the almighty in the grandeur of nature and a few (seemingly natural) ethical truths. Intellectually, it did away with the historical religions, Catholic or Protestant or Jewish, imprisoned in their orthodoxies and traditions, and unable to overcome their mutual hostilities. Their very diversity witnessed against them and their capacity to bring order to human life.

The Enlightenment is indeed the "Ellis Island" of modernity. Here the exiles from religious pluralism found their new faith and the new symbolism of order. Here is where the theoretical unity of truth as science is born, and where the effective realization of truth as technology begins, and where residual ethical and religious values, still embedded in European customs, prop up the one great Enlightenment value of the autonomous individual.

However, the Enlightenment prospect of a world unified by reason had some rude awakenings of its own in the 19th century and thereafter. The new Human Sciences re-interpreted the rational in relation to other forces – the unconscious, social structure, economic and political interests – which, in the end, seriously criticized reason's own naivete. The nature of autonomous reason, so thoroughly developed by Kant and Hegel, was made problematic by "philosophers of suspicion", Marx and

Nietzsche, who mounted a post-critical unmasking of reason. Marx attacked the unity of a bourgeois-rational world made secure at the price of the alienation of the masses. Nietzsche exposed the bad faith of the cultural amalgam still undergirded by Christian morals and a God the culture did not really believe in.

What Ricoeur has called the "hermeneutics of suspicion"[10] and MacIntyre 'unmasking"[8] becomes the post-Enlightenment style of reflection on reason. The end of the age of unifying metaphysics is the rise of multiple interpretations, of the conflict of interpretations, of the suspicion of autonomous reason itself. The Enlightenment's unhistorical understanding of reason, separated from and even antithetical to tradition, is vulnerable to this new philosophical-social critique.

Against this cultural backdrop, we can better understand Catholicism's reaction to pluralism. Regarded as a cultural outcast, still holding to what some took as outdated metaphysics, a supernaturalism, and mediaeval social world, Catholicism rejected modernity in turn and sought to preserve its own integrity doctrinally and socially. The otherness of modernity confronted the otherness of a mediaeval church. Against the new symbol of autonomous reason it asserted the need for faith, dogma, and historic revelation. Excommunicated by modernity as a serious intellectual faith, it excommunicated in return.

4. Apologetics to Engagement: Vatican I to Vatican II

From 1869 to 1965, there was a tremendous ferment in Catholicism on the subject of revelation and its related categories (e.g., dogma, scripture, authority). Until Vatican I, it was virtually unnecessary to define revelation since its reality (if not always the same interpretation of its contents) was universally accepted. With the Enlightenment, revelation became a radically problematic category for the rationalist culture. The decree on revelation at Vatican I called *Dei filius* tried to define the Catholic idea of revelation in opposition to certain 19th century philosophical trends – especially rationalism – which explicitly or implicitly denied it.

At the end of the 19th century, the Modernist Controversy again brought the issue of revelation to the forefront, this time in terms of the historicity of scripture and dogma. It attempted to apply the tools of higher criticism to the Bible and Tradition, but the effort was interpreted by Church authority as a rationalist reduction of doctrine and scripture. And one must admit that, despite the merits of the methods they tried to use, 19th century criticism was inclined to be reductive and rationalist. Except for a few voices, such as Blondel's[3], tradition as a category of religious authority was widely contested.[2] Church traditions and dogmas were regarded as the Hellenistic and Roman deformation of any earlier Jesus movement. Outside the world of romanticism, tradition in the culture was considered mere prejudice and anti-intellectualism.

In the course of the period from 1910–1960, the Catholic biblical movement improved its methodological approaches but maintained ideas of revelation and authority of a rather traditional Catholic kind. Gradually, the new methods were accepted by the magisterium, finally receiving their definitive approbation in the Constitution on Divine Revelation (*Dei Verbum*) at Vatican II. This document is much more comprehensive and sophisticated on the subject of revelation than Vatican I. Vatican II maintained the traditional concepts about revelation, but inserted a more biblical focus and description of its process of occurrence and contents. Nevertheless, the ecclesial corollaries of revelation in Catholicism – dogma, magisterial definition and authority, normative tradition, in short, the issues of traditional faith and critical method – still remain problematic topics. The aftermath of Vatican II demonstrates this clearly.

Vatican II opened Catholicism up to certain basic themes of modernity that have become the cultural standard since the Enlightenment. Among these are the tradition of human rights, religious liberty, historical criticism of scripture, ecumenical rapproachement, recognition of modern media and the benefits of the modern world in general. This was a remarkable achievement given the conflicts of the past four hundred years between Catholicism and Protestantism and the Enlightenment conception of religion and the post-Enlightenment's secularization of salvation. The council, in effect, accepted the new religio-cultural context as something it needed to understand and appreciate in order to carry out its mission. It identified itself with the "joys and the hopes" of the modern age, while still criticizing some of its excesses – particularly ethico-religious ones.

II. PLURALISM FROM THE PERSPECTIVE OF VATICAN II

Vatican II did not present a coherent vision of pluralism as such, if by vision we mean a consideration of this theme as a whole. It did, however, suggest certain motifs that express its attitude toward the phenomena of pluralism in modern society. Let us briefly indicate four such motifs.

1. Dialogue and Difference

The leading motif of Vatican II relative to cultural pluralism in general was the call for dialogue between the Church and all those parties with which she desires a relationship: other Christian Churches, other religious bodies, social institutions, governments, indeed, world culture. This call for dialogue was founded on the Church's basic willingness to accept the pluralisms of modernity as the actual point of departure for discussion and engagement. In some conciliar texts the very fact of pluralism is seen as a genuine cultural achievement. In others, however, pluralism symbolizes a crisis for modern culture. In both cases, though, the Church wishes to move beyond sterile polemics and strident apologetics to a genuine understanding of modernity's own self-understanding in all of its institutions. In order to assert a new kind of proclamation, Vatican II suspended the rhetoric of denunciation and moved toward an improved analysis of the post-Enlightenment world.

Dialogue implies a progressive conversation with the modern world and other faiths about the Gospel and its relation to the values, means, and ends of secular and ecumenical thought. Its hope is to discover some common ground either in human nature or in belief in one God or in Christ to forge the basis for further dialogue, clarification, understanding.The Extraordinary Synod of 1985 continued the process of assessing the situation of modern culture in relation to the Church's self-understanding and mission. The issues of inculturation, subsidiarity, and other concerns in the Final Report of the Synod make this clear.

2. The Autonomy of Realms

Vatican II asserted the fact and value of the relative autonomy of different realms of human experience and meaning. Religion, politics, science, art, social institutions should all enjoy a rightful autonomy or independence in order to realize their own distinctive ends without interference or undue limitation. Within the norms of the general moral law, based on the dignity of human nature and human freedom, autonomy in all such spheres of human activity is an essential value.

The council's idea of autonomy certainly presupposes a basic philosophy of human dignity and moral objectivity. It means a rightful independence from practical interference, but not schism between any realm of human action and moral values. It (a) supports ongoing dialogue among the various realms of human achievement, (b) asserts the right of the Church to address in an appropriate manner all moral issues, and (c) eschews any means other than reasoned public debate for achieving greater unanimity and cooperation among these various realms.

3. Christian Anthropology and Humanism

At the core of the Church's understanding of pluralism is its own anthropology based on revelation and issuing in the key idea of human dignity. This concept is the connecting link to all forms of modern humanism. It is the starting point for Vatican II's teaching on religious liberty. The nature and dignity of the human person is one of the key conciliar passageways from Christian revelation to secular culture. The idea of the person founds such central concepts as conscience, freedom, rights, and the meanings of birth, love, and death. The implicit wager of the council was that its shift to the key motif of the person, amply developed in modern Catholic theology, would make a thoroughgoing dialogue possible with modern culture. And why not? Since Descartes, Kant, and Kierkegaard, philosophic thought had made the anthropological turn fundamental for reflection. What this approach by way of the person did not see at the time was that major thinkers were questioning the vulgarized humanism of modern thought or advancing post-humanist ideologies.

4. Pluralism and Unity

In the vision of Vatican II, pluralism provides the new raw materials for the now much more complex task of discovering unity. In ecumenism, it takes the words of Jesus "that they all may be one" as its challenge. But, this inner-Christian teleology, for the Council, is also the hidden teleology of culture. It sees the fact of mass communications, for example, as the condition for realizing new kinds of understanding and cooperation among different peoples around the globe. Technology provides a new means for the human aspiration of unity and peace. At the same time, the Council also recognizes how technical progress can engender fresh separations and fragmentation in human experience. So, along with the Church's affirmations of pluralism and diversity, it keeps re-asserting a hope for greater unity. It adopts this as one of the major challenges for modernity, so rich in means, yet so often unsure regarding the ends for human life. In short, in the vision of Vatican II, pluralism enriches and elevates the essential task of unity to a more complex, but absolutely essential project. Pluralism is not an end, but a ferment of perspectives and structures calling for a new vision of human life in the world as a symbolically unified project.

These general motifs on pluralism *ad extra* at Vatican II need to be complemented by a properly theological discussion of the dialectics of pluralism and unity within revelation and theology. My focus will be to treat these dialectics briefly from the standpoint of the key concept of revelation since this is the fundamental warrant for the Church's message and mission.

The concept of divine revelation is the basic theme and warrant for theology. This concept tries to unify what are really a quite diverse body of historical experiences and texts, asserting that in them God has communicated " a message" or, since Vatican II, communicated "himself". In fact, we have many witnesses and forms of testimony to God within Scripture and within later Tradition. The Church's faith in the fact and purpose of divine revelation enables it to discern an Author and Source for the many witnesses to God it claims as its own. Revelation is then a meta-biblical catagory of unity in the sense in which the Church uses it. But this unity must be constantly reconquered by synthesizing the pluralism within the sources.

Within Scripture itself, we find many traditions in the Pentateuch and beyond this in the prophetic traditions, the wisdom traditions, four gospels, the Pauline corpus, and other New Testament writings. These express a pluralism of traditions immanent to the primary documents of revelation. This pluralism of traditions can be narrowed down even more from traditions to books (redactions) to passages, until the unity of the Bible is harder and harder to define. As a result of this, there is a new discussion going on today under the title of "canonical criticism" with the aim of reconstituting a unified concept of the Bible.

In another way, the various literary genres of the Bible problematize a unified picture of the process of revelation itself. This process is usually visualized on the basis of the prophetic genre characterized by the oracle, "Thus, says the Lord," or "The Lord spoke to me and said" which tends to fix our notion of revelation in every case as a divine voice speaking through a human voice. But, if we turn to the Torah and to Biblical narrative, God appears essentially as one actor in an unfolding drama (an actor, to be sure, who speaks), but not as a voice telling the narrator what to write. That image is borrowed from the prophetic experience and transposed onto the historical books. And, when we consider the wisdom literature, much of which is presented as the reflections of a sage on human experience, the idea of revelation changes again. For here, we often do not even have a meditation on the founding events of Israel, but philosophical observations on life, suffering, ambiguity, and death. Thus, the various literary genres of the Bible present different ways of conceiving divine revelation (a voice or dream, a story, tested maxims about living). Seeing these altogether requires an act of theological construction beyond the level of the texts themselves. That construction points to a theology of revelation as a unifying concept.

There is yet another way to see the dialectics of pluralism and unity in revelation. Biblical faith is based on certain foundational events that shaped the faith of Israel and the Church (Exodus, Exile, Jesus' Death and Resurrection). Nevertheless, biblical faith also absorbed religious imagery and themes from the surrounding culture. This is especially apparent in the Old Testament symbolism of God, covenant, kingdom, evil, and salvation. Oriental and Hellenistic influences on revelation – whether by supplying categories or paradigms or symbols for God and humanity – signifies that an extra-biblical religious experience survives within the Judeo-Christian faith. Religious pluralism, in the form of extra-biblical religion symbols, is both retained and reinterpreted within the biblical texts.

These first three forms of pluralism – multiple biblical traditions, multiple literary genres, and multiple religious sources in the Bible – express an inner-biblical pluralism that is related to the quest for the overall unity of the Bible. To supplement this, we can now turn to some theological categories guiding Catholic thought on the dialectics of pluralism and unity.

In the New Testament, the figure of Jesus Christ is, of course, the unifying element. And yet, even here, the primary testimony about Jesus is reflected through four gospels, each with a distinctive theological and communal interest. Beneath these interests, exegesis has been able to piece together a picture of the "historical Jesus" whose message was the imminent arrival of the Kingdom of God. With His Death and Resurrection, Jesus as the messenger gives way to Jesus as the"message". Thus arises the trajectory of Christologies, usually related to key titles given to Jesus in Christian communities. The variety of Christological titles in the New Testament witnesses to the diversity of ways of understanding Jesus from a "low" Christology of the prophet to the "high" Christology of the Logos or Son of God. In John's gospel, we find one of the highest New Testament Christologies expressed in the image of the Incarnation. The idea of Incarnation not only governs a Christology, but more generally characterizes the Christian faith as such. Christian Faith in the "Word who became flesh" is a faith with a built-in orientation to the material and the human. This grounds a whole theology of the Church, sacraments, and mission. It logically should repel any gnostic or docetic impulses in the Christian religion. The faith in the Incarnation, because it is related to belief in creation and sin, allows Christianity to incorporate finitude as a positive theme. Pluralism in cultures is one aspect of human finitude and therefore an invitation to the Incarnational principle in Christianity to adopt and revise cultural values for religious self-expression.

In another way, the Catholic notion of Mystery opens the way to a theology of pluralism. Mystery means that God's being and mind are infinitely beyond our human capacity to understand fully. So, while revelation informs us about God, it also instructs us in the finitude of our understanding of the divine. Our knowledge of God is true, but limited, and its very truth remains infinitely open to further comprehension. As a result, different attributes of God and divine acts – especially those which seem to pull in opposite directions – need not be opposed or mutually exclusive. To put it another way, the paradoxical character of God-language is the reflection of a human experience of the divine which shatters mutually-exclusive categories. Revelation itself is one such category for it seems to shatter itself on the distinction between the infinite and the finite. Yet, Catholic theology preserves it in a strong theology of language and analogical knowledge.

The idea of Tradition also includes pluralistic elements, since it is built up through the development and disintegration of multiple traditions. Tradition describes the gradual process of discerning the meaning of Christ, under the influence of the Holy Spirit, in the course of Christian history. This process works itself out in the whole life of the church though it is often most visible on public occasions such as councils, theological works, and papal pronouncements. Tradition is stimulated by historically contingent issues and challenges which help to disclose new meanings for revelation or new ways of presenting it. There is no *a priori* guarantee that at the moment such challenges arise any of the possible responses are adequate to preserving what has gone before, while meeting the new issue. We are dealing instead with a subtle, differentiated process of testing, adapting and rejecting the prevailing thought-forms to search out again and again the truth of Christ in relation to other forms of human knowledge and experience.

Above all it is the Catholic conjunction of Faith and Reason in theology and practice that connects Catholicism to pluralism in culture. This makes the Catholic mind suspicious of all forms of rationalism and fideism. So, it is opposed in principle to sectarian fundamentalisms and secular rationalisms as adequate assumptions about faith and knowledge. The Catholic spirit searches instead for those views of reality and those epistemologies that allow for a dialogical engagement of religious experience/language with empirical/scientific meaning. It discovers examples of them in thinkers who, though working from different philosophical points of view, are open to the phenomenon of faith, and in its own theologians who know how to keep the Catholic tradition faithfully and hermeneutically open to new questions and problems.

The inner pluralism of Catholicism which we have sketched out theologically through analyzing the notions of Incarnation, Mystery, Tradition, Faith and Reason is dialectically related to Catholicism's essential quest for unity. The union of the divine and human in Christ was the major constructive theological task of the early councils. Mystery does not lead to agnosticism, but to a theology of the indefinite fruitfulness of the Christian symbols. Tradition and Scripture must be seen and interpreted together. So also with Faith and Reason.

Beyond this, the Catholic teaching on the magisterium seems to suggest that faith must be taught and understood as much as possible as a unified doctrine. The role of authoritative interpretation of revelation is precisely to prevent the fragmentation of revelation. It is a principle of unity in a dialectical relation of service to the pluralism of theologies of the faith.

III. CONTEMPORARY PLURALISM AND CATHOLIC THOUGHT

Pope Paul VI called pluralism an "obscure term" because it covers so many phenomena that are rather loosely related. The term is used in virtually all human sciences – in philosophy, political theory, the social sciences – in different ways. However, it always signifies a liberation of thought from rigid orthodoxies or imposed doctrines. As such, it is the sign of intellectual freedom and ferment within all the human sciences, including theology. Let us now reflect on the phenomena of contemporary pluralism in philosophy and in what the sociology of knowledge calls society's " plausibility structures", to see their meaning for Catholic thought.

Contemporary philosophy not only offers a variety of views of reality, truth and value, but also has no single conception of itself as a discipline. So, it is not simply a matter of different answers to common questions, but something much more radical. There are at least four different ways of viewing what philosophy is all about ([10], pp. xvii–xxi) (1) Philosophy may be seen as an attempt to develop a *systematic representation of reality* that could unify our understanding of both the natural and social worlds. This view of philosophy is reflected in systems such as Thomism, Hegelianism, Marxism, and Scientific Positivism. (2) At the complete opposite extreme from this rather ambitious concept is philosophy seen as a pure *analytic of language* studying the logic and use of language in assertion and argument. In this approach, philosophy cannot offer any representation of the world beyond that of common sense or science. It stands guard, rather, over philosophy's pretension to metaphysical speculation. (3) Another view of philosophy would see it as an *exploration of human subjectivity* (existence), somewhat less sweeping than the systems and more speculative than analysis. Its subject matter is human experience, especially freedom, embodiment, and interpersonal relations. (4) Finally, philosophy may see itself as a *meditation on the end of metaphysics* as we have known it, along with a critique of subjectivity as a substitute for the breakdowns of metaphysics. In this case, philosophy seeks to articulate a sense of Being in more poetic and mystical categories.

In short, "present day philosophy offers conflicting views of its own mission ([10], p. xvii). It is not only pluralistic in the answers it gives to certain questions, but even in whether or not it deems certain questions meaningful at all or worth asking, never mind if they are possibly answerable. Each view of philosophy, as well as the different schools within each of the four types we have listed, seems to constitute a separate realm of categorization incommensurable with the other views.

Since Philosophy has been the primary organon or instrument for Christian theology as far back as Augustine, a crisis in philosophical thought such as we have described is bound to create problem for theology. In the present situation, theology has coped with this dilemma in several ways. First, since the turn to subjectivity, theology has adopted transcendental and existential philsophies to structure its own anthropology. A second approach has been to work with the newer hermeneutic types of philosophical reflection (philosophy as interpretation). Still another style confines itself to biblical, historical, or traditional (e.g, Thomistic) categories. As a result, theology's own self-understanding is made more complex and diverse. It more and more resembles the scattered condition of philosophical discourse.

Two Roman Catholic philosophically-minded theologians who address the conditions of pluralism as they relate to theological reflection are Bernard Lonergan and David Tracy. Lonergan approaches the phenomenon of pluralism using a transcendental method. He describes a basic "pluralism of expression" in religion that results from the various "differentiations" in consciousness and human activity.

Lonergan holds that there are three sources of pluralism: (1) linguistic and cultural differences that give rise to varieties of common sense; (2) specialized modes of apprehension and language to deal with specific realms of human meaning (e.g., art, scholarship, philosophy, etc.); and (3) the degree of intellectual, moral, or religious conversion that a person may experience ([5], p. 326).

For the first source of pluralism – culture – Lonergan uses the distinction between a normative view of culture (viz. the "classicist" view) and an empirical view. The classical type is a combination of Western, Hellenistic, Roman, European cultural models. Empirical types, on the other hand, are not normative, but pluralistic and horizontal. Catholicism has adopted the resources of classical culture to express itself for over a thousand years. Now it must meet the challenge of speaking to and in the categories of a variety of cultures and modes of common sense. For Lonergan, however, this is a challenge for the *communication* of doctrine, not for its plausibility as such. For Lonergan, the real threat to the unity of faith lies not in the attempt to embody Catholic faith in different modes of common sense or in various forms of technical expertise. It lies rather in "the absence of . . . conversion" ([5], p. 330). This may take at least three forms: (1) absence of conversion in those who govern or teach in the church; (2) the movement of Christianity out of classicist and into modern culture; and (3) the lack of understanding of or appreciation for the various forms of differentiated consciousness as they affect faith ([5] p. 330). So, problems of authority, truth, cultural change, and forms of knowing radically threaten the unity and integrity of faith.

David Tracy ([12], pp. 3–46) deals with pluralism in terms of three types of "publics" or social contexts for theology: the church, the academy, society at large. The task of theology, precisely because it is concerned with questions of ultimate truth and goodness, is to make explicit the plausibility structures and the claims to truth that are present in these three major spheres of human meaning. It serves each of these "publics" by working for greater clarity in making the claims and counter-claims to truth and goodness understood. It seeks to advance the cause of truth by evaluating the "adequacy" of the truth-claims made by these various publics. In short, theology strives to be a discourse related to distinct publics (and their relative plausibility structures) and a way of assuring that the issues of ultimate concern in religion will remain public issues of meaning and truth.

Tracy makes "publicness" a main task of theology in order (a) to offset the fragmentation of the self in pluralistic societies in which meaning is distributed across various realms of action, and (b) to counteract the privatization of truth and value so convenient for a religiously pluralistic culture. Theology is the endeavor to keep various realms of meaning, various structures of plausibility, and the various "selves" of the modern individual in contact with each other.

Both Lonergan and Tracy develop complex theological methods for identifying the different forms of pluralism and for making theology equal to the task of articulating Christian truth in the context of diverse cultures and diverse plausibility structures. They aspire to a kind of "higher unity" of faith and culture so characteristic of Catholicism. But, they don't pretend to do anything more than provide some maps or procedures for carrying out this task. They see it as a task radically affected by our modern situation of cultural pluralism and philosophical-social pluralism. The theologian must reenact within his own consciousness, sympathetically and critically, the outlooks and ways of speaking of the audiences he wishes to address. Whether we use the transcendental method of Lonergan or the correlation method of Tracy, theology is struggling for the grounds that could sustain a dialogue between Christianity and modern culture. One might say they are trying to develop the conditions for the possibility of a genuine dialogue between the Church and human culture.

Without having answered the problem of the collapse of metaphysics, or the historicist reduction of the transcendental claims of revelation, contemporary theology finds itself drawing increasingly on the sociology of knowledge to understand its cultural situation. One of the basic insights of this sociology is that human beings tend to accept as credible (or plausible) what others in their own social group believe. Consensus – especially about matters that go beyond direct sense experience – makes certain convictions plausible. The lack of consensus and modes of social reinforcements of beliefs will render such beliefs less plausible. Whatever intrinsically supports or diminishes the credibility of certain beliefs also determines to a large extent whether those beliefs can be seriously entertained as options for people. The generic level of praise, disapproval, indifference that a society has toward certain practices – independent of the moral arguments people might make about them – is clearly one of the major determinants of human attitudes about those practices. Usually in somewhat closed or traditional societies, the plausibility structures we've been talking about reinforce the basic social beliefs at virtually every point. When societies become pluralistic in religious and ethical belief, plausibility is thereby reduced for all those sectarian convictions. Plausibility is maintained, though, for those social beliefs still necessary for the functioning of the pluralistic culture. In such a world, the various minority faiths will invent forms of social reinforcement for their own beliefs since society at large will no longer provide them. Thus, in a pluralistic society, religious groups need to understand the plausibility structures which either confirm or disconfirm their own systems of belief. They also need to develop belief-maintenance structures of their own in response to the cultural position.

As H. Richard Niebuhr has shown in *Christ and Culture*[9], the modes of response to the surrounding cultural posture by the church may vary from opposition to accommodation to synthesis to tension to coversion. No typical response is completely adequate because the cultural potentials and problems change in relation to Christian faith. Protestant liberal theology, for example, judged that the plausibility structures of modern society would not really permit people to accept a Christianity focused on the supernatural. As a result, Christianity was given an existential, ethical, and psychological re-interpretation. Catholic neo-conservative theology tries to avoid the liberal adjustment of faith's *credenda* by bolstering its structures of authority and the specificity of doctrinal commitment. Whether any article in the system of belief counts as a 'disposable believable" (to use Ricoeur's term) because it is merely a religious residue from a previous culture or whether it is a "genuine inference" from divine revelation is the most fundamental question for all theology working in a pluralistic culture. At the extreme, a pluralistic ethos tends to reduce all faith statements simply to cultural expressions, possibly meaningful to those who belong to that culture (the religious sub-culture). Religious truth claims look, from a certain pluralistic perspective, like the last form of cultural chauvinism.

Catholic theologians who endorse the American ideal of pluralism, because of its many benefits for the pursuit of truth, understand it as the condition for an open cultural conversation about truth and human values. This conversation goes on between the poles of complete relativism and intense sectarian convictions. In between these two extremes are the more or less plausible positions that can gain a larger public hearing. As long as a lively public debate goes on about moral issues, today mostly in the form of public policy debates, there will be an ongoing correction in the depth of moral consensus. When this debate itself is criticized, if religious convictions lie behind some moral stances, if substantive values beyond the right to privacy, individual freedom, and tolerance are seen as an imposition of the alien ethical beliefs of a minority on the society, pluralism leads back to relativism and the privatization of truth and value.

In a pluralistic society such as ours, it is imperative that religious institutions lead the way in the public conversation about values. That may well be one of their essential social functions in a pluralistic ethos. Tocqueville called religion one of our fundamental political institutions in the sense that it helped to cultivate the values and virtues necessary for the right exercise of freedom ([11], p. 292). Now, instead of cultivating values naturally accepted by members of a society, religion finds itself trying to argue for the very plausibility of those values in an ethically-pluralistic culture.

Alasdair MacIntyre ([7], p. 199) has observed that in our culture, precisely because of the pluralism of moral perspectives we have been talking about and diverse underlying assumptions about values, some moral questions are "systematically unsettleable". He argues that because of moral pluralism the conditions for general moral consensus are lacking, and their absence precludes the establishment of strong traditions, which in turn means that authority cannot be sustained, and the rational practices that "require the recognition of authority" seem arbitrary ([7], p. 204).

MacIntyre has attempted to uncover the cultural implications of pluralism by indicating how our structures of moral debate and rationality are affected by it. He sees a kind of chain reaction in the post-Enlightenment world moving from the breakdown of traditions to the loss of authority to the feeling that moral practices themselves are quite arbitrary. Explaining this dilemma is the deeper predicament, according to MacIntyre, arising when moral argumentation must deal with the phenomenon of incommensurability. When moral assumptions are radically opposed to each other, the various conclusions, corollaries, and inferences from moral reasoning cannot find enough common ground to communicate with one another. Just as Americans do not suppose that one can settle religious disagreements by rational apologetics and debate (they regard religious convictions as the result of being raised in a certain religious tradition or feeling comfortable with a certain teaching), so moral convictions – due to the pluralism of moral opinion – can only have a subjective or emotive plausibility. Thus, MacIntyre sees no way out for a pluralistic culture but to elaborate in social policy some kind of co-existence of different moral perspectives based on individual freedom and rights. Within that culture, those who still maintain an ethics based on moral tradition will exist in minority communities. He even recommends to these individuals the task of finding new ways to "rescue and recreate authority within communities that will break with the pluralistic ethos" ([7], p. 212). He concludes his book *After Virtue* with the same appeal to create "forms of community" within which a moral tradition and a vision of what constitutes virtue can still be maintained ([8], p. 263).

MacIntyre's view of the implications of pluralism on morality has been taken up recently by H. Tristram Engelhardt, who sees ethics being divided between a public, secularist ethic suitable for a pluralistic society and particularist ethics espoused by individuals or religious communities. The secularist ethic basically comes down to an ethic of personal freedom without any corresponding objective scale of values. In order to maintain peace and order in such a society, according to Engelhardt, we have to turn from substantive moral objectivities to procedures for assuring freedom for individuals to exercise their own preferred moral style ([4], pp. 1–53). Both MacIntyre and Engelhardt have located the question of pluralism in the tension between personal conviction and public order. To what extent can a society remain coherent and be morally pluralistic? How deep can this schism go?

To understand this problem, I offer the following observation. Looking back over the history of morality, I perceive three great schisms within ethics. The Book of Job introduced the first schism in the separation of the physical order of misfortune from the moral order of culpability ([10], pp. 90–91). The price that was paid for this was that much suffering had to become absurd in order that sin might become truly a spiritual matter. The second schism concerns the splitting apart of morality and religion in the modern period. Values could be justified only by reason instead of by a religious tradition. The price paid for this separation was the loss of a religious horizon or structure of motivation for morality in general. The third schism is the one we are discussing here, namely, the dissociation of morality from social structures of reinforcement and plausibility. Morality then becomes private. The price for this separation is that conscience must live simultaneously in two ethical worlds each immoral by the standards of the other.

What is the Catholic response to these successive disjunctions of ethics? What is it to say to a pluralistic vision of morality? It has theoretically accepted the schism of Job between the moral and the natural order. It has accepted natural law to support its ethical stances. Now, it is faced with the effects of the schism of public and private morality. It has certainly tried to maintain itself, as MacIntyre recommends, as a community of moral consensus in post-Enlightenment culture. It has relied on a strong sense of tradition and the rational practices flowing out of them. It has attempted to resist the breakdown of religion into reason, and reason into mere competing interests or plausibility structures. It did not yield ground to modern ideologies of the state or technology. Slowly, and somewhat unevenly, it moved from an apologetics of refutation of modernity to selective appropriations of Enlightenment achievements (historical criticism, religious liberty, etc.) according to its own principles. Yet its own vision of unity in faith and the unity of faith and reason makes Catholicism sit somewhat awkwardly in a pluralistic ethos. It affirms some aspects of pluralism – e.g., human autonomy, religious liberty, and ecumenism – and still hopes for the reunion of Christian bodies and the drawing closer of all religious bodies. It challenges other pluralistic phenomena – e.g., relativism, atheism, and secularism – for reducing the nature of human existence to purely worldly or material dimensions. Meanwhile, within its own institutional structures, the Church has begun exercising greater authority by the magisterium over theological speculation to limit the extent of dissent, seeing to the unity of faith. More specifically, it seems to interpret some forms of pluralism as just yielding to the plausibility structures of modernity in areas such as church order, sexuality, and economic ideology.

Daniel Bell ([2], p. 347) has defined religion as serving the unity ot culture in terms of the answers it provides for the existential aporias of human life – birth, self, freedom, destiny, death. At one time, the Catholic vision provided such a unity for culture. Pluralistic culture does not look to a singular religious institution to provide a comparable unity today. Instead, it relies on the principles of individualism and rights in public while exploring in private messianic ideologies, fundamentalistic faiths, mysticisms, and traditional religions as the keys to the existential questions and the order of historical meaning. Catholicism must reflect on all these as well in the attempt to discover the grounds in revelation for a shared vision of humanity and its historic destiny.

To sum up, we have reviewed some typical variations in Catholic history on the theme of pluralism and unity, and reflected on the tension of unity and pluralism within Vatican II, revelation, and doctrine. In the final section, we discussed the deeper philosophical and epistemological problems associated with modern pluralism along with the responses of some representative Catholic theologians.

The Catholic vision of pluralism is certainly a more open one today, but it is also mixed, tentative, and provisional. Catholicism continues to work for realizations of a certain unity of faith and existence. It does this by virtue of its public mission and public self-identity as the Church of Christ. Above all, it does it by fostering the plausibility of religious faith and higher moral sensibility in whatever culture it finds itself. How – not whether – to do this, is the fundamental ecclesial issue of our time. It is the ecclesial equivalent of the ongoing social task of building a meaningful order of human existence in terms of "ends divine and human" ([14], p. ix).

St. Mary's Seminary and University
Baltimore, Maryland,
U.S.A.

NOTES

[1] Max Weber was the first to analyze the Church/Sect distinction.

[2] Blondel's major work, *Action* (1983), establishes the theoretical foundations for a practical ontology of life as the condition for a constructive theology of tradition.

BIBLIOGRAPHY

1. Bauer, W.: 1971, *Orthodoxy and Heresy in Earliest Christianity*, Philadelphia Seminar on Christian Origins (trans.), R. Kraft and G. Krodel (eds.), Fortress Press, Philadelphia.
2. Bell, D.: 1980, *The Winding Passage: Essays and Sociological Journeys, 1960–1980*, Basic Books, New York.
3. Blondel, M.: 1956, *Lettre Sur les Exigences de la Pensée Contemporaine en Matière d'Apologetique* and *Histoire et Dogme*, Presses Universitaires de France, Paris.
4. Hauerwas, S.: 1986, *Suffering Presence: Theological Reflections on Medicine, the Mentally Handicapped, and the Church*, Notre Dame, Notre Dame, Indiana.
5. Lonergan, B.: 1972, *Method in Theology*, Herder and Herder, New York.
6. MacIntyre, A.: 1969, *The Religious Significance of Atheism*, Columbia Univesity Press, New York.
7. MacIntyre, A.: 1975, 'Patients as Agents', in S. Spicker and H. T. Englehardt, Jr. (eds.), *Philosophical Medical Ethics: Its Nature and Significance*, Kluwer Academic Publishers, Dordrecht, Holland.
8. MacIntyre, A.: 1984, *After Virtue* (2nd Edition), Notre Dame, Notre Dame, Indiana.
9. Niebuhr, H. R.: 1951, *Christ and Culture*, Harper and Row, New York.
10. Ricoeur, P.: 1978, *Main Trends in Philosophy*, Holmes and Meier, New York.
11. Toqueville, Alexis de: 1969, *Democracy in America*, trans. G. Lawrence, J. P. Mayer (ed.), Doubleday-Anchor Books, New York.
12. Tracy, D.: 1981, *The Analogical Imagination: Christian Theology and the Culture of Pluralism*, Crossroads Books, New York.
13. Troeltsch, E.: 1931, *The Social Teaching of the Christian Church*, trans. O. Wyons, George Allen and Unwin Publishers, London.
14. Voegelin, E.: 1956, *Order and History, Vol. 1: Israel and Revelation*, Louisiana State University Press, Baton Rouge.

JOHN H. WRIGHT

A BRIEF COMMENTARY ON "NOTES ON A CATHOLIC VISION OF PLURALISM"

May I first of all express my appreciation for Father Leavitt's rich and wide-ranging paper. As a systematic theologian I am grateful for his historical survey of Catholic attitudes toward pluralism, for his assessment of the grounds provided by Vatican II for affirming and qualifying pluralism, and for his view of contemporary pluralism and Catholic thought.

I was reminded in the course of studying his paper of an observation made by Arthur Koestler in *The Roots of Coincidence* to the effect that every subordinate unity within a larger unity must for its own survival and well-being be intent on two seemingly opposed projects. First, it must seek to affirm and preserve its own distinctive identity; and second, it must relate in openness to the larger unity that surrounds it. If it concentrates only on itself, then it perishes in atrophy and isolation. If it concentrates only on openness to the world around it, it perishes by being dissolved into the environment ([1], pp. 111–114). Catholic concern for pluralism must make sure that pluralism within the Church strengthen rather than diminish its distinctive identity, and that pluralism outside the Church both enrich the Church and be enriched by it rather than fragment it or evoke a premature rejection.

The main point of focus of Father Leavitt's paper was on the pluralism in the world around us and how we as Catholics should relate to it. This question was raised more particularly within the framework of the ethical questions raised by issues of human health and the practice of medicine. It is not the precise concrete questions with their varied and diverse answers that we wish to consider here, but rather how we can relate to this pluralism in a way that both maintains Catholic identity and at the same time learns from the pluralism around us and makes a contribution to it.

I wish first to make some observations and raise some questions about Father Leavitt's paper, and then to offer some reflections on pluralism which may focus our consideration in view of the previous essays.

I should like to emphasize the word "quest" in Father Leavitt's initial description of the Catholic approach: "a quest for unity and order in

255

Edmund D. Pellegrino et al. (eds.), Catholic Perspectives on Medical Morals, pp. 255–260.

revelation and reason" ([2], p. 231). We do not claim to have it all together. Whatever our achievements, we need to press on, to seek for unity and order between faith and religious understanding on the one hand, and reason and secular culture on the other. There is a tension here, but the Catholic approach resolutely refuses to relax the tension either by opting for fideism or for rationalism. For this reason I would add to the Catholic witnesses about concern for unity and order (such as Augustine's vision of historical meaning, Thomas's synthesis of reason and faith, and various conciliar restructurings and proclamations) humble research and enquiry after truth.

I am inclined to see the initial emergence of "early Catholicism" at some point prior to the Pastoral Letters. Someone once remarked when asked where this tendency to structure and organize religious faith and practice was first observable in the Bible, "I think you first see it in the Yahwist (the J tradition in the composition of the Pentateuch)." This may be pushing it back somewhat too far, but as a phenomenon among Christians, it seems to me observable in Paul's concern for offices and for order in 1 *Cor* 12:27–28; 14:26–33 and *Rom* 12:4–8, in his trip to Jerusalem to present his preaching to Cephas and others "lest somehow I should be running or had run in vain" (*Gal* 2:2), in his addressing the "bishops and deacons" of the Church at Philippi (*Phil* 1:1), in the description of the Church as "built upon the foundation of the apostles and prophets" (*Eph* 2:20), and in the many memories of the early life of the Church preserved in the Acts of the Apostles. Christian assemblies and churches never appear as simply gatherings of free charismatics, where each one does whatever he may be inclined to do with no one in charge.

Among the significant influences on the intellectual life of the Middle Ages I would add Boethius' *On the Consolation of Philosophy*. This was a work that was read and studied by every educated person in Europe for a thousand years. It presented a way of looking at all the opinions that were abroad concerning the real meaning of life and how we should live in the light of that meaning. It can, I believe, be enlightening for us today.

I am inclined to see the pluralism of the Scholastics in the Middle Ages as more relevant to our contemporary concerns than does Father Leavitt. The various schools of theology grew up in response to outside influences on the Christian understanding of the faith. Plato, Aristotle, Plotinus, Averroes and others provided tools for understanding and communicating the meaning of Christian revelation. This produced great variety in theology, a genuine pluralism that was fruitful at times, though it may have ended in a certain sterility, as much from the lack of intellectual greatness of the thinkers as from the principles from which they proceeded.

In looking at the great thinkers of the Middle Ages, I think we have to see in Aquinas something more than one who brought together *lectio divina* and the philosophy of Aristotle, thereby enabling him to deal with incommensurable terms like nature and grace. We have to reckon with the fact that Aquinas' own mind was stimulated to an enormously significant original contribution of his own, his metaphysical analysis of being in terms of potency and act going beyond Aristotle's matter and form. This was epochal in the history of human understanding and even today no one can deal intelligently with the problem of being and neglect Thomas's insight. It also enabled him to deal creatively with the pluralism of his culture, distinguishing, evaluating, and uniting from the higher viewpoint this insight gave him.

It has sometimes occurred to me that the conflicts between different points of view at the time of the Reformation and of the 17th century scientific controversies were as much a matter of personalities as of principles. After reading John Todd's most recent biography of Martin Luther[4], I had the strong impression: it didn't have to happen that way. The theological differences were not so deep as to have to divide the Church; if Rome had taken Luther more seriously, dealt with him more fairly, been less concerned about the loss of revenue coming from the sale of indulgences, and if he had been still more patient, less a prey to melancholy, and able to bend a bit more without sacrificing any principle, then Lutheranism could have been a school of spirituality and theology within the Catholic Church, rather than a separate communion institutionally divided from it. Likewise, the Galileo case could have turned out differently with neither science nor religion weaker for it, if Galileo had been less inclined to interpret scripture and the Holy Office more open to the evidence of the telescope and mathematics. But the personalities involved would not let it happen that way.

I would like to suggest that the basic attitude of the Enlightenment in seeking unity through the use of reason continues to be valid, in spite of the naive optimism of that movement and its failure to recognize the many circumstances that condition and hinder the operation of reason. It is the "pure desire to know" that Bernard Lonergan speaks of, which by its dynamism offers us hope to get beyond polemics and controversy to the unity of mutual understanding and agreement. Such unity cannot be legislated into existence by some authoritative decree.

I think that Father Leavitt is right in speaking of pluralism in the meaning of "revelation". The model at Vatican I, echoing the model of prophetic speech, seems to have been divinely communicated propositions expressing mysteries of the godhead otherwise unknowable. Faith accepts these propositions. But revelation must be understood as prior to propositions, as the divine self-communication given in experience to one who receives it in faith and who then may more or less successfully articulate it in propositions of various sorts: prophetic, interpretive, poetic, narrative, parabolic, and even fictional. However, it was not clear to me why Father Leavitt says that revelation is a category which "seems to shatter itself on the distinction between the infinite and the finite" ([2], p. 243).

Let me now develop briefly a few thoughts on pluralism as it affects our concerns. Pluralism is sometimes used to mean the same as diversity or variety, and thus merely points to a fact. In a more evaluative sense pluralism means an acceptance and even welcoming of this diversity and variety. The Extraordinary Synod of 1985 made a further distinction in discussing pluriformity and pluralism: "The one and unique Catholic Church exists in and through the particular Churches ([3], pp. 2–3). Here we have the true theological principle of variety and pluriformity in unity, but it is necessary to distinguish pluriformity from pure pluralism. When pluriformity is true richness and carries with it fullness, this is true catholicity. The pluralism of fundamentally opposed positions instead leads to dissolution, destruction and the loss of identity" ([3], N. 44, p. 7).

One's attitude toward pluralism is very much a function of one's epistemology. If it seems to you that the truth is clear and obvious and is there for any one to affirm who has eyes to see, then pluralism can seem only a compromise with error and ignorance. However, if it seems to you that truth, especially religious and moral truth, is ultimately mysterious, and that only through much enquiry, reflection, discussion, and cooperation do we move toward less and less provisional and partial statements of the truth, then pluralism will seem to you an indispensable condition for the discovery and exploration of truth.

It is often said, somewhat along the lines of the Extraordinary Synod, that we can welcome complementary pluralism but not contradictory pluralism. Complementary pluralism seems to be very close to the pluriformity spoken of by the Extraordinary Synod. It is simply opposed to uniformity, and sees the same meaning expressed in different ways; or different but complementary meanings are sustained by different individuals. Contradictory pluralism means accepting as good a condition in which opposed or contradictory positions are set forth. Does contradictory pluralism always lead to dissolution, destruction, and the loss of identity, as the Synod said? It seems to me that if the contradictory pluralism represents final statements of what is deemed to be the truth, then it does lead to dissolution. But the path toward a definitive statement of truth is frequently as much dialectical as logical. It is only as the merits of opposed positions are freely debated and discussed that the truth can gradually emerge, to be clarified, nuanced, and finally affirmed fully and unhesitatingly. An attempt to avoid this process of debate by using authority to suppress a position does not ultimately serve the truth, but simply promotes ideology, while diminishing the credibility of those who proceed in this way. It is not the motives but the wisdom of this way of proceeding that is chiefly in question.

Catholicism does of its own special genius seek to promote unity and universality out of the great variety of opinions and positions and attitudes that are abroad in the world. But this unity can never be simply decreed into being; it is the fruit of honest effort and enquiry; it means trying to understand and to represent fairly other positions, acknowledging the truth that one seems to find there, while pointing out in a reasoned way what seems to be false or mistaken. We must endeavor to promote a sense of cooperation among all disciplines in seeking for the truth, more willing to affirm than to deny, recognizing as Leibniz once observed that most people are right in what they affirm and wrong in what they deny. It means for us believers seeking and welcoming the light of the Holy Spirit, and acknowledging the special ministry of the Church's Teaching Office in this process of enquiry, a ministry that is not equipped to answer all questions single handedly, but to make sure that in our enquiries we do not overlook indispensable values and perspectives. For the Pope and the Bishops have no other source of knowledge than the rest of the faithful: sacred scripture, the documents of Tradition, and the *sensus fidelium*. It is only after carefully pondering all this that they can credibly express the faith of the Church and the Spirit that is moving the Church, without confusing this with their own tastes and opinions. History offers many examples of this sort of confusion.

I would like to close by underlining two sentences from Fr. Leavitt's paper: "In the vision of Vatican II, pluralism enriches and elevates the essential task on unity to a more complex, but absolutely essential project. Pluralism is not an end, but a ferment of perspectives and structures calling for a new vision of human life in the world as a unified project" ([2], p. 241).

Gonzaga University
Spokane, Washington
U.S.A.

BIBLIOGRAPHY

1. Koestler, A.: 1972, *The Roots of Coincidence*, Hutchinson, London.
2. Leavitt, R. F., S. J.: 1988, 'Notes on a Catholic Vision of Pluralism', in this volume, pp. 231–253.
3. Synodus Episcoporum, Bulletin: Dec. 9, 1985, Rome.
4. Todd, J. M.: 1982, *Luther, A Life*, Crossroad, New York.

PART VI

AGAPEISTIC MEDICAL ETHICS

ROBERT SOKOLOWSKI

THE ART AND SCIENCE OF MEDICINE

Just as a mathematician is most fully himself when he is calculating, so a physician is most fully himself, as physician, when he is engaged in medical activity. Medical activity is the actuality of medicine, and both the art and the science are to be defined and understood in relation to it. The art and the science both are as potential activity. It would be a distortion to regard medicine as, say, essentially a science, essentially an understanding of certain natures and relationships, something to which applications were accidental; or to consider it as an art that could be itself without ever coming out of hiding, without becoming active. Both the science and the art would be out of focus without the activity.

But although medical activity is the climax of medicine, it is only a part of medicine. No medical activity activates the whole of the art and science. Knowing what to disregard and what not to do, and being able to do other things in other circumstances, are necessary as a background for acting here and now as a physician. A physician is always more than what he does at any moment. Indeed, the partiality of the practice extends even farther, for the single physician is himself never the embodiment of the entire art and science of medicine. He knows that there are other doctors who know and can do things that he does not know and cannot do. Medical science and the medical art, medical reason and medical ability are essentially distributed and only partially realized in any agent and in any transaction. They remain largely latent even when they are activated in the climax of an action. They are always largely a background and a disposition.

Consequently, the concern of the physician is not only to carry out properly the transaction he is engaged in, but also to preserve the being of his art and science, as the dispositions out of which his actions as a physician emerge. He is obliged not only to his patient, but also to his art and science.

I

Let us consider medical activity more closely. A medical action is a transaction between physician and patient. The term "patient" is some-

263

Edmund D. Pellegrino et al. (eds.), Catholic Perspectives on Medical Morals, pp. 263–275.
© *1989 Kluwer Academic Publishers. Printed in the Netherlands.*

thing of a misnomer. It connotes total passivity and suggests that the target of the action is not thoughtfully involved in the exchange. On rare occasions this may occur; someone may be brought unconscious and with a broken leg to a physician, who may simply proceed to set the broken bone. But this is obviously not normal. The normal situation is that the medical reason and medical ability of the physician are shared to some extent with the patient. The patient has to comprehend to some degree what is going on, and he also has to share in the medical activity. He has to take the medicine, perform the exercises, change his diet, and do many other things that could be called medical care. He has to take care of his health. But his understanding and his activity are under the sway of those of the physician.

Because the reason of both physician and patient are engaged in medical activity, such activity has some of the features of a conversation or dialogue. A few years ago, a psychiatrist named Stanley A. Leavy published a book called *The Psychoanalytic Dialogue*[3]. In it he tries to show that the psychoanalytic relationship is like a revival of human exchanges rather than like the relationship between a patient who is suffering a disorder and a scientific observer, the physician, who discloses or diagnoses "the objectively present infection or tumor" ([3], p. 16). Leavy is correct in stressing the interpretative and conversational character of psychoanalysis, but the understanding of the medical relationship he uses as a foil is not adequate to medical activity. The physician does more than reveal a disorder. From the beginning, since the patient comes to the physician and begins by explaining how he understands his own condition, the medical exchange is one in which the patient also thinks and acts.

But of course the patient and the doctor do not think and act in the same way. Because of his art and science, because of the background behind what he is actually doing now, the physician has an authority in this relationship that the patient does not enjoy. His authority is analogous to that of a teacher toward his students. His assessment of the situation and of what is to be performed is the governing assessment as long as the medical relationship lasts. The patient's reason and ability only participate in those of the physician, but they do indeed participate. Because he too is a rational being, the patient is not like an animal that is brought – that does not bring itself – to the veterinarian. It is the presence of reason in *both* patient and physician that makes the medical relationship a human transaction and not merely the exercise of a technical skill.

But what makes it into a specifically medical exchange? How is the physician different from the electrician who comes to wire my house, the mechanic to whom I bring my car, or the barber who cuts my hair? The difference is that in the medical exchange *I myself* am at issue in a way in which I am not in the other cases. Sometimes the medical is defined by reference to the bodily and its health and disorders. But we should begin with something more basic. The medical is special not because my *body* is at issue, but because *I* am at issue. The medical is so prominent in life because it is engaged in my concern for my very self: not for my house or my car, nor my looks simply, but for my own being. We are human not just because we can calculate, but also and more originally because we are an issue for ourselves, and the medical exchange does strike at this concern. To take the most dramatic case, when we are waiting for the results of a critical test, one, let us say, that is to disclose whether or not we have a serious illness, we find ourselves to be at stake not just in what we have but in what we are. Whether I will continue to have a future, whether what I have done and undergone in the past will still remain as a springboard for further action, how my being in the world will be modulated, all this changes from being a silent undercurrent in life to being the issue I am directly concerned with. The medical exchange is an echo of our anticipation of death, and the prominence that death gives to my very self and what is my own also arises, in a muted form, when a serious medical issue must be faced. And even in the less urgent medical cases, even when we are trying simply to improve our health or to treat some illness that is merely inconvenient, we are acting in relation to ourselves and not merely to our possessions.

Of course my concern for my health is not the summit of my concern for myself. It would be hypochondria or neurotic obsession to gear my life primarily to health. I am much more an issue for myself when, say, I am about to act loyally toward my friend or gratefully toward my benefactor or generously toward someone in need; or perhaps when I am about to betray my friend, steal something, or destroy someone's career. When I act morally I will not only affect someone else, but determine myself as well. I will be as these actions are. Whenever any particular issue surfaces for me, I also surface as an issue for myself. These actions will be mine and will be me in a more intimate way than my bodily well-being will be mine and me. And yet, nevertheless, my bodily illness or health will also be me in another way. I am concerned with myself and not just with my body when my bodily wholeness is an issue.

To observe that we as human beings are essentially an issue for ourselves is not to say that we are somehow selfish or self-centered in a morally reprehensible way. To speak about our self-concern is not a moralistic observation but a remark about the formal structure of a human person, the structure that underlies our ability to say "I" and "mine" and "my own" ([1] § 41). Even the most generous act of self-sacrifice and self-forgetfulness is "mine" for the agent. It is owned and is considered his very own. That is why it can be noble and admirable, not just a characteristic of the species but an accomplishment of this person.

II

Now this highlighting, in the medical context, of my concern for myself, this drawing of that concern into the foreground occurs not in solitude but with the involvement of the physician. I do not just worry by myself that I might be getting sick. I become an issue for myself in this explicit way in the company and the presence of the doctor. And the doctor is there not formally as a friend, but as a professional. He is there as the embodiment of the medical art and science, not primarily as, say, my brother or sister or colleague. This does not mean that he becomes malevolent or indifferent, but it does mean that his benevolence is not the same as that of a friend. He recognizes my being as a person, but, as a representative of the art and science of medicine, as one who professes that art and science, he would treat anyone else in my situation as he is treating me. His art lifts him out of a personal involvement: he is there not as John Smith but as the doctor, and I am there not as Aloysius but as the patient.

But the doctor is also John Smith and the patient is also Aloysius. We cannot separate these aspects of being, but we must distinguish them, and we must acknowledge how each accompanies the other and how we can shift from one to the other.

The physician is the embodiment of the medical art. To possess the art of medicine is not merely to know a lot of functional relationships and to know how to operate with them. An art is not merely a technique. An art is the ability to accomplish a good. An appreciation of the good to be done is programmed into the art, programmed not as one more step in its procedures but as saturating every step from beginning to end. Medicine as technique is what is left over when the medical art is drained of its sense of the good. Techniques are the finger exercises of medicine. Technique is cleverness or dexterity, the kind of thing in which computerized expert systems can help us; it is exemplified by, say, accuracy in correlating infections with combinations of antibiotics, or by precision and quickness in surgery. Such techniques can be rehearsed and they can be admired as techniques, but they are abstractions. They are not done in and for themselves; they belong to a larger and different kind of whole. The art itself is not the sum of techniques, and the doctor does not present himself as the embodiment of many techniques. As one who professes the art, as a professional, he presents himself as someone who will both know and be able to bring about the medical good that is possible in the situations that are brought to him. We expect the doctor to be benevolent not as a friend, but as an embodiment of the art, as someone who seeks the good of the art.

But what is the good of the medical art? It is the health of an individual or community; more specifically, it is the preservation or the restoration of the health of an individual or community. And this good exists in a complex way.

Health is there before there is a medical art. The good of medicine is not like the good of football or music, a good that comes into being as the sport or skill comes into being. Health is first of all a condition of a living thing, and it is first and foremost a good for living things, even before there is medicine. Furthermore, health is not only sought by things that are alive, it is also restored by them through the self-healing that living things achieve even without art. The medical art comes on the scene when something can be done about health, when health appears not only as a desirable condition nor just as the outcome of the natural process of healing but as a good that can be brought about through deliberate action. The art becomes developed as we become able to do more and more about health. Such growth takes centuries of experience, insight, and practice, and the art then becomes so enlarged that only specialists can learn it. Health becomes the object of a profession. However, health as a good to be achieved by medical practice, health as the good of the medical art, is also the same health that was sought as a good before there was medicine. The one good becomes targeted or wanted as good from two directions, from the point of view of the living organism that wants to be healthy, and from the point of view of the medical art; in this criss-cross, the perspective of the medical art is secondary and derivative. The physician heals because health already is a good for the living being. The good of the healing art is always already a good for the one who wants to be healthy.

Because health, while remaining the same good, is a good from two different perspectives, there can be interesting oscillations between the two ways in which it can be taken. An innocent example of this oscillation occurs when, say, a patient who has undergone a successful operation is exhibited to a troop of medical students, who are to admire and learn from the skill that was exercised by the surgeon. The good of the art becomes paramount for the moment, and for the moment the art becomes appreciated almost as technique. The patient as such, and his view of the good that was achieved, fade into the background; he may even be a bit embarrassed by all this attention. No harm is done, of course, but the shift in perspective is illuminating. It shows that the good can be seen from two angles and that in a small way the priority of the patient's view of the good can be muted.

In a much more radical conflict of goods, the doctor may use his skill to torture or to take revenge on someone. He wants not the good of his "patient", but the bad or the harm of his target; the harm to the person's health becomes formally the doctor's good. In such a case the doctor exercises his medical technique but not his art; indeed he violates his art, contradicting and acting against the good that is programmed into it. Strictly speaking, he does not act as a physician when he does such things; he struggles explicitly against his art and practice.

And in still another possibility, the physician may become, say, avaricious or ambitious, and he may perform his actions primarily in order to enrich or advance himself. In this case the art keeps insisting in his actions; the good of the art is not directly violated, and it may indeed be promoted if the practitioner is to achieve the purposes he has set. He will make less money if he fails to cure his patients. Still, the good of the art has become a means rather than an end. It is still a good for him, but good primarily as a means to wealth or power, no longer an end in itself. In such a case there are not two but three perspectives on the medical good: that of the patient who wants to be healthy, that of the art, and that of being a means to the doctor's advancement. The doctor's medical activity becomes subject to alien pressures, to a field of force generated by a purpose that is only accidentally blended with the medical good, and it becomes possible that the medical good itself will not be sought in a particular action, but that it will be distorted in view of the purpose it is made to serve. The doctor will not explicitly want the bad of the patient, as, say, the torturer does, nor will he fail through carelessness, as a poor practitioner might, but he can allow the harmony between the good as wanted by the patient and the good as sought by the art to be disharmonized by the purposes he introduces into his assessment of what is to be done.

In the best case, however, there is no distortion, and in acting according to his art the physician also seeks the good of the patient. Because the art of medicine aims at something that is a good for the patient, the doctor, in the exercise of his art, seeks the medical good of the patient as his own good. He pursues, professionally, what is good for another. He does this as a practitioner of his art. The nature of his art, with the perspectives it provides on the medical good, gives the physician this harmony, and it makes him, in the good exercise of his art, not only a good doctor but also essentially a good moral agent, one who seeks the good of another formally as his own.[1] The doctor's profession essentially makes him a good man, provided he is true to his art and follows its insistence.

It is important to realize that by being faithful to his art and to the good that is programmed into it, the physician is helped to become a good man. In the philosophical analysis of virtue, a distinction is usually drawn between being good in the various arts and being good as a human being. In the dialogues of Plato, for example, we are often reminded that someone can excel as a craftsman but fail in the art of living. This is a legitimate distinction, but we must not turn it into a necessary separation. Rather, in the important arts, and specifically in the art of medicine, the goodness of the art will help shape the character of the person who practices it, because it will form him into someone who seeks the good of another as his own good. It will help the physician to be excellent not just as a doctor but as a human being. The physician can therefore be grateful to his art not only for the skills he learns from it, but also for the character that it bestows on him.

In addition to his professional benevolence, the physician may also wish his patient well as a friend or as a colleague, but this is not the first relationship between them. Their friendship does not establish them as doctor and patient, and another person who is not a friend or a colleague would not be somehow deprived in entering into a medical relationship with the doctor. The art, with its network of perspectives on the good in question, establishes the way the doctor and patient are to be related in the medical transaction.

III

We have discussed the medical good as seen by the patient and as seen by the doctor representing his art. But the good is not simply what people desire; sometimes we desire what only seems to be good but really is not. We must ask about the truth of the good in question, and in asking about truth we approach medicine as a science and not just as an art. We approach medicine as an understanding.

Just as the medical art cannot be fragmented into a sum of techniques, so medical science cannot be broken down into information about functional relationships. It cannot be broken down into knowledge about the workings of the various elements, powers, and processes that are relevant to human health. We in our time and culture are inclined to break it down into such particles, because the science that was introduced by Francis Bacon, Descartes, and Galileo does tend to see the world as a network of relationships and laws. The science we inherit from these authors looks away from the forms of things to the elements and forces out of which the things are made. But if this is all our science is supposed to know, then the art associated with it will be mere technique. If all we know are a lot of "if-then's", then all we will be able to do is to manipulate these relationships, to pull *these* levers and let *that* happen. If we are to go beyond this, we must understand what health is, and since health is an excellence of the human being, we must understand what a human being is and how the excellence of health is related to his complete human excellence.

In fact, every physician knows that he is seeking a good in his practice of the medical art, and he has some sense of how it is related to the over-all human good. But given the modern understanding of science, the physician will not be comfortable in considering this knowledge to be part of his science of medicine. He will say he knows it rather through common sense or through his religious belief or his cultural values. But surely this reflects a deficiency in the science of medicine. The modern science of medicine has been enormously successful in the treatment of diseases and in understanding the mechanisms of the body. Perhaps it could not have achieved its successes had it not limited its focus to the study of structures and functional relationships, had it not abstracted from the definition of larger wholes and forms. Perhaps the controversies and the imprecisions that arise when these issues are faced would have impeded the progress that was made within the more constricted horizon. But although this restriction can be understood and may need to be tolerated, it certainly is not something with which we should be satisfied.

Medicine as science still has more to accomplish along the paths it has been following in the past few centuries. There are diseases still to be understood and treated, there are bodily structures and functions still to be explained. But a more complete natural science would involve not just more of what has been done, but also an understanding of a different kind: the definition of health and disease, the clarification of how the medical good can be related to the political, economic, and personal good, the understanding of the relationship between health and human agency (an issue that is especially urgent in the psychological branches of medicine). To give a more specific instance, medical neurophysiology is incomplete so long as we are not able to say, scientifically, what it means for certain neurological activities to be not just chemical or electrical processes, but also the act of seeing a tree or hearing a song; or what it means for another neurological phenomenon to be identifiable with an immune response. The full scientific truth of neuroanatomy is not achieved until the sense and excellence, the form, of the neurophysiological is understood, until we can state the sense and excellence of what the neurophysiological does; and since medicine aims at preserving or restoring a healthy state, it too is incomplete as a science, as an understanding of the good it is to accomplish, until these dimensions have been introduced. The more complete medical science must reintroduce an appreciation of form, but it must do so in a way that is commensurate with the great advances made in medicine and biology during the era in which science was silent about form.

IV

Let us now turn to the religious aspect of medicine. The medical practitioner knows that he is not the origin and controller of everything that goes on through his art. He knows that he only assists processes and forces that occur on their own initiative, that healing originally takes place quite apart from the mind and agency of the physician. The medical art itself is mysterious: its discoveries, its exercise, its successes, are never mastered by any one person, and they depend on forces that will never be fully comprehended and ruled. There is much to revere in medicine.

In the Hippocratic Oath this reverence is expressed in the invocation of Apollo under the name of Apollo Physician – reason and art in the medical form – and in the invocation of Asclepius, the founder of medicine who became a god, and of his daughters Hygieia and Panacea, and finally of all the gods and goddesses. The Hippocratic physician acknowledged his obligations to the powers to which he and his art were subject ([2], ch. 9).

But for Christians, the world has been purged of such divinities and powers. The world is understood as created by the one God, who could be all that he is, in undiminished perfection, even if he had not created. For the pagans the world appears as the necessary, unquestionable background for all the natures and events that come to be within it; the world is a context even for the gods, who take their place within it; but for Christians the world itself is no longer the final setting. The world itself is profiled against a deeper horizon, one in which there could possibly be only the divine and nothing else. The world, with everything in it, does exist, but it exists out of the possibility of not being; and its being is not through inexorable necessity but through the unforced generosity and choice of the creator.[2] And within the created world, human beings are understood not only as created but also as redeemed by God and allowed by grace to share in the divine life. Human being, furthermore, is elevated within the world because it becomes the place where, in the incarnation, God becomes united with a part of the world and presents himself as part of the world. The divine nature and the person of the Logos become united with the human nature in Christ.

Are the medical science, art, and activity affected by being placed in this new context? How are they affected? It would not be appropriate to define the changes primarily as new or different moral obligations introduced by Christian belief. Rather, any new moral emphases are themselves the outcome of a new understanding of what is involved in medicine. Christian faith reveals further the truth of medicine and what medicine does.

But what further truth is there to be stated? We have already described the nobility of the medical art, its capacity to make its practitioner good not only as a doctor but also as a man. What else is there to say?

The Christian understanding of the world is in some tension with modern science, including modern medical science, in regard to the recognition of the forms or natures of things. Christian belief would emphasize the reality of form as against the scientific tendency to stress the elements and forces out of which things are made. In regard to human being, it would emphasize human nature and the virtues and obligations inscribed in human nature. Of course one need not be Christian to acknowledge form; one need not be Christian to appreciate that things have their specific excellences and internal ends, and to appreciate that the human good and the human end are indicated by human nature. But this sense of a human end is heightened in Christian belief, because living according to virtue becomes not only noble and proper to man, but also in accordance with the creative will of God. The light shed by creation does not change our human goods, but it gives them a deeper sense. But when Christian theology deals with the new context of creation, it must be careful not to turn the noble into the obligatory; it must introduce its new context in such a way that the noble remains intrinsically excellent and does not become simply a duty placed on us by a commandment.

In regard specifically to the medical art, it is not the case that Christian belief would have a special access to the virtues and obligations associated with medicine, that it would be able somehow to derive medical goods from new moral principles that are not available to those who are not Christian. It does not have a short cut to the ethics of medicine. Rather, theology must first work with the excellence of the medical art as it shows up to those who practice it well; it must recognize and elaborate the internal good of medicine; then it must place this good in the context of creation and redemption and show how the integrity of the good is sustained and how it is heightened and highlighted by Christian belief. It is a question of an identity that is maintained within a new and distinctive set of differences, an instance of the preservation and perfection of nature within the context of grace.

One thing that is appreciated differently in Christian belief is the place of personality in the world. In a strictly evolutionary understanding of the world, the human person is taken as emerging out of the impersonal, subhuman forces of nature; the personal is a kind of excrescence, a fortuitous appearance in nature. In a more classical pagan understanding, the personal may be taken as a permanent part of the world: besides the material elements of the world, there were gods, spirits of various sorts, and even, for some thinkers, the human species as eternal. The world was not conceived to be without mind and life. But in the understanding presented by Christianity, the personal is exalted to a much higher dignity. The personal does not evolve out of the impersonal, nor does it contend with the impersonal, but it could, in God, be all that there is. The very existence of the world is the effect of a personal action. The personal is therefore seen in a new perspective, and even the human person is elevated to a new condition. The existence of the human person is seen as the effect of a personal act, and the destiny of the human person, as redeemed by Christ, takes on a cosmic significance.

And what is at issue in medical activity is not just a living organism, but we ourselves in our bodily and psychological being. If we understand ourselves as created and redeemed, then the issue in medicine is also understood differently: not that the natural good of health is changed, but that its sense and emphasis are modified. Human life and the transmission of human life, for example, are appreciated as a mystery even in the course of nature, but they are more deeply reverenced when life is seen as given by the creator and redeemed in Christ.

The healing of human life was one of the ways in which the incarnation of the Son of God manifested itself and one of the ways in which the presence of the Holy Spirit was shown in the early church. The bodily and psychological cures in the Gospels and the Acts of the Apostles may well serve as an indication that the act of healing is one of the first of the human activities that can be informed by grace. Medicine can then not only express the knowledge of science, the skill of art, and the benevolence of human action; it can be an expression of charity as well.[3]

Catholic University of America
Washington, D.C.,
U.S.A.

NOTES

[1] On the form of moral transactions, see R. Sokolowski [5].
[2] On creation and its role in Christian theology, see R. Sokolowski [4].

[3] I am grateful to John A. Feeley and Thomas Prufer for comments on an earlier draft of this paper.

BIBLIOGRAPHY

1. Heidegger, M.: 1962, *Being and Time*, trans. John Macquarrie and Edward Robinson, Harper and Row, New York.
2. Kass, L. R.: 1985, *Toward a More Natural Science*, Free Press, New York.
3. Leavy, S. A.: 1980, *The Psychoanalytic Dialogue*, Yale University Press, New Haven.
4. Sokolowski, R.: 1982, *The God of Faith and Reason*, University of Notre Dame Press, Notre Dame.
5. Sokolowski, R.: 1985, *Moral Action: A Phenomenological Study*, Indiana University Press, Bloomington.

٢

EDMUND D. PELLEGRINO

AGAPE AND ETHICS:
SOME REFLECTIONS ON MEDICAL MORALS FROM A
CATHOLIC CHRISTIAN PERSPECTIVE*

INTRODUCTION

A Catholic "perspective" on morals – medical or otherwise – implies a coherent view of the moral life that transcends purely philosophical ethics in some distinctive ways. What those ways consist in, whether they are different in kind or degree, and whether they entail a distinctive moral life are all important and still problematic questions.

At a minimum, any definition of a Catholic perspective must confront some of these fundamental questions: Does a Christian belief entail a content and methodology distinct from, and closed to, philosophical ethics? Is uncritical fideism or unrelenting rationalism the only alternative? Is ethics as a reasoned discipline reconcilable with ethics as a response to the moral imperatives of the Gospels or their authoritative interpretation by the Official Church?

Many of these issues are engaged throughout this volume and in contemporary theological discourse. They form the inescapable backdrop for this essay, which examines them in the limited confines of professional ethics. My principal aim is to examine the way in which the central virtue of the Christian life – the virtue of Charity – shapes the whole of medical morals.

This essay proceeds in two steps: In the first, it examines some of the conceptual relationships between reason and Charity in traditional and contemporary ethical discourse. In the second, it tries to show how Charity shapes our interpretations of: (1) the central principles of philosophical ethics – beneficence, autonomy, and justice; (2) the way we construe the healing relationship itself; and (3) the way we make some of the crucial moral choices facing health professionals and society today.

Two important caveats must be established at the outset:

First, the Christian virtue of Charity is the central distinguishing feature of a Catholic and Christian perspective on the moral life. Christian ethics is by definition therefore an "agapeistic" ethics. But this is not synonymous with the situation ethics of Joseph Fletcher [11]

277

Edmund D. Pellegrino et al. (eds.), Catholic Perspectives on Medical Morals, pp. 277–300.
© *1989 Kluwer Academic Publishers. Printed in the Netherlands.*

which claims also to be agapeistic. Fletcher's form of ethics eschews principle and precept and dissociates reason and Charity in ways incompatible with a Catholic perspective.

Second, to put some emphasis on the practical consequences of an agapeistic ethics does not necessarily place orthopraxy and orthodoxy in opposition to each other[45]. Rather it is the special task of Catholic and Christian ethics to reconcile doctrine and practice and to effect a harmonious equilibrium between them. Only in that equilibrium can the moral life of the Catholic Christian become the integrated whole required by both reason and faith.

The theses I wish to argue are three:

First, any medical moral philosophy ought to begin with the nature of medicine itself, as a human activity, which, on grounds of natural reason alone, imposes certain obligations on the physician and other health professionals. This is the "internal" morality of medicine, itself[25].

Second, a "Catholic perspective" begins but does not end with this internal morality. It adds dimensions of insight and obligation that grow out of Christian teachings of Charity such that charity "informs" ethical reasoning in certain specific ways. Just how charity and reason relate is a central question for traditional and contemporary accounts of the moral life. A "Catholic perspective" for our times should try to reconcile the traditional and contemporary views.

Third, though the relationships between reason and Charity and between traditional and contemporary views of Christian ethics are still under discussion, certain moral choices are clearly more consistent with the virtue of Charity than others. It *does* make a practical difference if one professes to be a Christian as well as a physician, nurse or administrator.

In this discussion, three topics pertinent to the Catholic perspective are consciously excluded: (1) the relationships between ethics as a reasoned discipline and the teachings of the official magisterium; (2) the place of the casuistic method in medical moral decisions; and (3) the moral dilemmas of reproductive and sexual morality. This first is amply covered in papers by Hughes, McCormick, Fuchs, and Cahill in this volume; the second is the subject of a recent study [23]; and the third is so much at the focus of Catholic moral theology today that it needs no further discussion here ([29, 7]).

Instead, my focus will be on the kind of person the Catholic physician and health professional should be. This is more consistent with a virtue-based ethic than with a focus on the solution of specific moral quandaries. Though the emphasis is on virtue-based ethics, the great importance of linking virtue, principle, obligation, and rule is clearly acknowledged. It is beyond the scope of this essay to forge those links except tangentially.

I. THE INTERNAL MORALITY OF MEDICINE

Many prominent thinkers[1] conclude that belief in God, Creation, Redemption, and the Incarnation does not provide specific answers to general or medical moral problems. They argue that the whole of the right and the good is open to human reason since the good is intrinsic to the world God created. On this view, God asks obedience to moral law because it is good; moral law is not good simply because God asks obedience to it.

In the case of medical ethics, Fuchs summarizes this argument as follows: Medicine is a moral enterprise because it deals with human problems. The ethics of medicine derives from medicine as a human activity. Its moral nature is prior to, or at least not dependent on, faith. Medical ethics thus must accord with human understanding, and in this sense it has a certain autonomy[14].

Fr. Sokolowski follows essentially the same line of argument. He begins with the phenomenology of medicine as a special kind of human activity[51]. He focuses on the art of medicine and the way it functions in the physician-patient relationship. What is at stake is the person of the patient. He and the physician, as rational beings, each play a part in effecting the end of medicine which is the good of the patient. In this relationship, the physician is the embodiment of the medical art whose end is the patient's good.

On this view, beneficence is a moral obligation which is programmed into the art. If the physician does harm, he violates the art. If he is faithful to the art, he becomes a good moral agent and is, himself, enobled. Thus, the art establishes the way in which physician and patient should relate to each other. This is the "internal" morality of medicine itself, and it is derivable by reason with Christian belief. I have argued similarly for the obligations that are derivable from the fact of illness, the vulnerability of the patient and the physician's promise to help[36].

According to Sokolowski, Christian belief affects this internal morality of medicine in several ways: it reveals more fully and truthfully the good intrinsic to the art; it corrects the tendency to reductionism and the neglect of form characteristic of modern scientific medicine; it highlights the dignity of the human persons – doctor and patient – who confront each other in the healing relationship.

Christian belief thus expands the range of insights available to ethics as a reasoned discipline.[2] As Lisa Cahill points out[6], the Catholic perspective operates best when it brings together philosophical reflection, religious images, logical interpretation, concrete human experience, and magisterial teachings. Each contributes to a fuller comprehension of the natural law and enables Christians to make moral choices in conformity with the spirit of Gospel teachings.

On this view, ethics as a reasoned discipline becomes insufficient to express the whole of the moral life. It becomes more than the application of principles and rules. It gains insights from the Gospel teachings on Charity that are not admissible into ethical discourse by those who reject those teachings.

Romano Guardini, in his meditation on the Sermon on the Mount, puts it this way:

Once we restrict ethics to its modern sense of principles it no longer adequately covers the Sermon on the Mount. What Jesus revealed there was no mere ethical code but a whole new existence, one in which an ethos is immediately evident. . . . Only in love is fulfillment of the ethical possible. Love is the New Testament ([17], p. 79).

and Thomas Merton says similarly:

We must of course point out that mere ethics, as a moral philosophy has its limitations. It needs to be completed by a higher science that apprehends other and more mysterious norms which have been revealed to man by God and which arise out of the deep personal relation of man to God in saving grace by which man is oriented to his true and perfect finality his ultimate fulfillment as a person in the love of God and of his fellow man in God ([31], p. 127).

II. THE PHILOSOPHICAL STATUS OF CHARITY-BASED ETHICS

We agree with Guardini and Merton, that we must go beyond "mere ethics" to fulfill the Christian commandment of love. Like St. Paul (*Col.* 2,8), we must be wary of the possibility of the submergence of Charity by philosophy. Both Guardini and Merton recognize that it is the complementarity between faith and reason that distinguishes the Catholic moral tradition and preserves it against the heart-over-head experiential ethics of Jansenists, Quietists, and Modernists. Nonetheless, a Charity-based agapeistic ethic poses difficult philosophical questions that remain problematic.

Frankena, for example, notes the philosophical dilemma in the double imperative of Christian ethics – to love God and one's neighbor as oneself ([12], p. 58). He argues that an agapeistic ethic could be grounded in the principle of love of neighbor but that love of God could not be derived from beneficence alone. Without faith in God's existence we could not derive the command to love God. But for the Christian this is the ground of his agapeistic ethic. The Christian loves God because God has created all that is good. Charitable beneficence is grounded in God's love for us and in His revelation of that love. It follows from faith, which is the "virtue of entry" into the Christian life and which assures us of a personal relationship of love with God.

Yet, Frankena admits ". . . . there is a sense in which the law of love underlies the entire moral law even if this cannot be derived from it" ([12], p. 57). That the law and all the prophets are summed up in the love of God and neighbor is not a conclusion of reason, but neither does it violate reason.

These metaethical difficulties do not preclude the possibility of a Christian moral philosophy. Every moral philosophy rests ultimately on some ordering principle, whether it be the categorical imperative, the principle of utility, moral sentiment, love of man without God, or love of man because of God. In each philosophy, there is, at the outset, an act of faith in some ordering principle. To deny that any such principle exists is itself an ordering principle. For the Christian, the existence of God and revelation is one such starting position. It does not, on that account, have a lesser claim to coherence than moral philosophies that deny both God and revelation.

In recent years, Catholic thinkers have approached the question of a specifically Christian and Catholic ethics with renewed interest. They have examined the ways in which Charity and reason are related by linking traditional moral theology to contemporary philosophy and psychology. These attempts are not always mutually reconcilable. But they do open up possibilities for the fresh synthesis of old and new ideas that a comprehensive Catholic moral philosophy requires today.

For Catholics, the central question is how to reconcile an ethics based in reason, principles, and precepts with the fact that the fullness of the Christian ethos of charitable love is somehow beyond ethics in Guardini and Merton's sense. Is it possible to avoid the extremes of an unthinking totally experiential fideism on the one hand, and a rigid, unfeeling, legalistic rationalism on the other?

Hallett attempts an answer by linking Charity and reason through analytic philosophy [18]. He focuses on the criteria by which a Christian ethic based in Charity can be judged. He posits a system of "Christian Moral Reasoning" that purports to reconcile the traditional allegiance to reason and objective norms with the concrete particulars involved in actual moral decisions. He proposes on the one hand to avoid both the absolutization of precepts he finds in traditional Christian ethics and the abandonment of objective norms in the agapeistic situation ethics of Joseph Fletcher.

Hallett thus argues for a "third position" between the extremes of preceptive and anti-preceptive ethics. This he calls "value ethics" which makes use of the insights of analytical philosophy and judges the Christian nature of an act or decision by a balance of Christian values over disvalues. Hallett's concept of Christian value maximization is provocative and deserves further examination. His proposal is, however, largely procedural. It by-passes the substantive metaethical issues. He tries to give due pre-eminence to Charity, even while denying the value of a specific hierarchy of values which would give the first place to Charity.

An even more ambitious attempt to re-define Christian ethics comes from the "Murray Group" – theologians thinking, in the spirit of John Courtney Murray, from a "North American" viewpoint. They examine Catholic theology "through the eyes of American philosophy"[33], by which they mean the philosophies of James, Dewey, Whitehead, Pierce, Royce, and especially Jonathan Edwards. From these sources they claim to derive a theology that is not only rational and objective but also takes into account the experiential, affective, aesthetic and pragmatic dimensions of the moral life. They hope thus to balance the excessive rationalism they perceive in traditional Catholic moral philosophy by drawing on experience, feeling, and imagination.

This perspective is well represented by William Spohn[52]. Spohn expands on the place of discernment, and *metanoia*, in making concrete ethical decisions by what he calls the "reasoning heart". The reasoning heart does not contravene reasoning and moral principles in ethics. Rather, it operates within them, but draws also on imagination and experience to illuminate individual moral decisions. "Discernment", says Spohn,

remains a personal search for the action of God in one's own history and in the events of the world. Although its conclusions are not morally generalizable as judgments of rationality are, the reasoning heart of the Christian finds normative guidance in the symbols and story of revelation ([52], p. 66).

On this view, what distinguishes Christian ethics is that it is motivated, in the moment of moral choice, by a specific set of affections – those that most closely correspond to the character, affections, and goodness of Christ. Christian ethics differs from purely philosophical ethics because it can draw on these affections which, without conversion, are closed to non-Christians.

This is not the place to enter into a detailed critique of this interesting mode of theologizing. Spohn recognizes the dangers of intuitionism and situationism in his approach. Just precisely how the balance is struck between moral sentiment and moral reason in this form of theology is not clear from Spohn's or from the other essays in the anthology. Like Hallett, Spohn is seeking a middle position between the extremes of fideism and rationalism. As with all middle positions, finding the precise point of balance is the crucial challenge.

Another linkage between Christian belief and philosophy is Pittenger's application of Whitehead's process philosophy to the understanding of religious affirmation [43]. Pittenger examines the origins of Christian faith in humans as a Whiteheadian "event" and how, in those terms, it apprehends the reality of Jesus and the way that reality opens "a window into God". Here, as with any interpretation of Christian belief in terms of a specific philosophical system, the links with more traditional ways of philosophizing need critical examination. With Whitehead, or any of the other "North American" philosophical approaches, there is always a danger of eroding the still-viable truths of the more traditional Catholic moral philosophy.

Pertinent to many of the attempts to balance affect and intellect in moral decisions are the current discussions on the relationships between moral cognition and moral motivation. Here the question is this: Does a recognition of the right and good impel to doing the right and good? Are these totally separable operations, and, if they are not separable, how are they in fact linked?

Such questions are relevant to the nature of Christian ethics and to the various meanings of discernment as it is found in St. Ignatius, Karl Rahner, or the Murray Group. Some theorists suggest that moral motivation is a function of one's sense of self, rather than a function of one's conceptualization skills on the one hand or a matter of anticipations of external bribes and rewards on the other[55]. Conversion to the Christian Faith creates St. Paul's "new man", one whose sense of self is shaped by the Christian virtues. Is this the way being a Christian makes a difference in ethics? Is it this transformation of the self that links cognition and motivation? Is this the locus for the "illumination" spoken of in the more classical moral theologians and the "discernment" preferred by their contemporary counterparts?

Each of these approaches attempts to reinterpret, in contemporary terms, the more traditional viewpoints on the relationships of philosophical and theological ethics as found in St. Thomas. They are interesting in their own right, and they force us to seek a deeper understanding of the Catholic moral tradition. Gilby reminds us that St. Thomas

Jogs us to remember that philosophical ethics is contained within *sacra doctrina*, that the discourse is drawing from sources beyond the reach of reason alone, and that from within there is a reaching out to a good beyond reasonable statement, the ultimate good, transcendent yet not abstract, the burden of all yearning which is God himself, the end beyond measure of all morality ([15], p. 148).

St. Thomas' extended treatment of the human psychology of habit and passion gives a personalist cast to his ethics that should be reassuring to contemporary thinkers who fear the rigidity and absolutization of an exclusively preceptive moral system. As Walgrave points out, St. Thomas was

in his own way a personalist . . . because he combined a radical methodical intellectualism with a personalistic anti-rationalism, showing the radical insufficiency of ratiocination in determining the principles and in deciding the practical issues of moral life ([54], p. 214).

In another recent series of studies Pinckaers shows how central to Thomistic ethics was the doctrine of charitable love, as revealed in the beatitudes. ([38, 39, 40]) Pinckaers traces the evolution and maturation of the idea of the good and its link with ethics from the *finis bonorum* of Cicero, through Augustine's dictum that the Sermon on the Mount provided the "perfect pattern of the Christian life" to St. Thomas' own morality. In them is to be found the sure guide to happiness sought by the pagans as by Christians. Pinckaers sees in Thomas' interpretation the possibility of once more reconciling morality and the desire for happiness, which he sees divorced in contemporary ethics.

Pinckaers also takes the view that duties and obligations are secondary in the morality of St. Thomas. They are the "crutches" of the virtues, placed at their service:

La Morale de St. Thomas est donc une morale du bonheur et des vertus, groupant celles – ci autour de la foi, de la charité et des vertus cardinales. Ainsi s'explique le peu de place accordé dans la Somme á l'obligation morale ([38], p. 109).

He goes on to show how during the Reformation and counter-Reformation, both Protestantism and Catholicism divorced morality and faith, the one in the direction of rejecting reason, the other in exalting the natural law. Both are misguided, says Pinckaers. If we wish to understand Thomistic ethics, we must restore the primary place to faith. The essential fact is the internal disposition to charitable action formed by faith and helped by the Holy Spirit. This, he holds, gives freedom, not arbitrariness, to Christian ethics.

In his reflection on the methodology of St. Thomas, T. C. O'Brien distinguishes St. Thomas' vision of beatitude from Aristotle's. For Aquinas beatitude is the result of union with God, not of virtuous activity, as in Aristotle. The moral quality of human acts is, thus, measured by the degree to which they advance Charity as the prime moral principle. Beatitude is a form of friendship, and therefore of mutual love between God and man – a relationship that comes only through Grace.

It is not, then, a question of seeing a natural moral structure, then filling it in by identifying the ultimate end as God; the vision of grace and charity is first; the moral structure is chosen to express something of its intelligibility ([32], p. 114).

St. Thomas thus reverses the usual order of a naturalistic ethic, in which practice of the virtues can move man to his proper end of happiness. Instead, for Aquinas the only way to the fullness of beatitude is through grace and Charity. Such fulfillment is for man "above the condition of his nature but not in disregard or negation of his nature" ([32], p. 97).

One of the more promising recent re-examinations of the philosophical foundations for Christian ethics is the interesting amalgam of Thomistic realism, Christian existentialism and phenomenological methodology that goes under the heading of "Lublinism"([24, 26, 56]). Here Christian ethics is grounded in the personalism of an ambitious philosophical anthropology. One of its most prominent exemplars is Karol Wojtyla, now John Paul II, whose personalist ethics centers on the lived experience and participation in love of the acting person [56]. On that view, the traditional values of Catholic morality are preserved and enriched by some of the creative ideas in contemporary European philosophy. This is in distinct contrast with both the analytic thrust of Anglo-American ethics and the legalistic bias of some of Catholic moral theology in the past.

These recent attempts to bridge the gap between ethics as an enterprise of reason and ethics as fulfillment of the law of love are valuable extrapolations of the traditional Catholic "perspective". They offer fresh insights into the links between philosophical and theological ethics, between contemporary and traditional ways of doing ethics, and between faith and reason, intellect and will, virtue and duties, agape and moral principles. We are far, however, from bridging all these gaps. To do so would require something of a "Summa Moralia Christiana". Pinckaers has recently offered an impressive new synthesis of the old and new sources of Christian morality that moves in this direction in an impressive way ([41, 42]).

In all of this, it is important to avoid an overly eager acceptance of the new or an overly rigid adulation of the past. Intuitionism, situationism, psychologism, and biologism are easy traps to fall into. Yet, the divisions between preceptive and non-preceptive ethics may not be as wide as their respective protagonists may feel. Those who favor existential and experiential modes of ethical thinking need a deeper and updated reacquaintance with Thomistic ethics. Those who favor the more traditional modes need to acquaint themselves with the richness of possible connections between Thomas' thought and contemporary philosophy and psychology.

The great strength of St. Augustine and St. Thomas was their capacity to engage in creative dialogue with the dominant cultural ideas of their times. Their intellects, illuminated by Faith, were able to apprehend what was congruent with the law of Charity and to discard what was not. If we can still philosophize and theologize in the spirit of Augustine and Thomas, it seems possible that a truly comprehensive Catholic medical moral philosophy will emerge. Such a philosophy would tell us more about the kinds of persons we ought to be than the rules we ought to follow. In medical morals it would call for Catholic health professionals who possess an intellectual grasp of moral principles as well as a capacity to apply them in the spirit of charity. In this way they might fulfill that law of love which, as Guardini says, "*is* the New Testament" ([17], p. 79) and without which Christian ethics is not possible.

III. CHARITY, THE FORM OF THE VIRTUES, AND THE CATHOLIC PERSPECTIVE

Whatever new synthesis may be effected by the dialogue between traditional and contemporary Catholic perspectives Charity remains the ordering principle.([3,20,48,49]). D. J. B. Hawkins puts it well:

What is distinctive of the moral teaching of the gospel is not a new code of morality or a new theory of its basis, but the insistence on raising morality to the level of love. The commandments are to find their full meaning and completion in the love of God and our neighbor for God's sake ([20], p. 28).

This is what St. Thomas teaches when he makes Charity the "form" of the virtues (*Summa Theologiae*, 2a2ae,Q23,art.8). Gilleman provides a particularly compelling account of Thomistic teaching on this point. He shows how Charity gives every virtuous act and, indeed, every virtue, a supernatural moral worth by orienting them to their final end, which is a supernatural one, i.e., union with God ([16], p. 53).

For the Christian, therefore, the moral life is conducted from several perspectives of creatureliness and incarnation ([20], p. 16). When we recognize that we are creatures, we see that everything we possess is held in trust for the Creator's purposes, not just our own. When we accept the fact of the Incarnation, we recognize that God's revelation of Himself to us makes it possible for us to lead a supernatural as well as a natural life. We are enabled to go " . . . beyond the possibilities of nature left to itself"([20], pp. 16–17).

Clearly, in an agapeistic ethic, the motivation for being moral is explicitly different from what it is in a naturalistic ethic. The Christian knows that doing the right and the good is a means of growing closer to God the creator and Redeemer. He also has in Charity a light that illuminates the central dilemma of philosophical ethics, i.e., why some rules and principles are morally imperative and others are not.

Fulfilling moral obligations is, in a sense, a way of encountering God.

As John Crosby has argued, in that encounter, we conform to God's will and recognize that " . . . the binding force of moral obligation ultimately derives from divine command"([8], p. 317).

This entails more than fulfilling duty in response to a reasoned argument about what ought to be done. Instead, the encounter with God in moral choice demands that the end of our reasoned judgment must be a right attitude of mind and heart. The virtue of Charity, therefore, consists in disposing moral judgments to their right end. It fuses the qualities of both mind and heart, of reason and faith – a fusion without meaning in a non-agapeistic ethic.

IV. MEDICAL PRACTICE AND THE VIRTUE OF CHARITY

If there is something distinctive about the Catholic perspective, it ought to be manifest in practical moral decisions. Since medicine is a praxis, an activity with its own internal goal, that goal – the good of the patient – is a moral one. A Christian perspective should therefore dispose the physician to decisions that would be, in the sense discussed above, formed by Charity. Or, to put it another way, as Fr. Sokolowski suggests, it would transform healing into an act of grace.

To be able to make choices consistent with the virtue of Charity requires a particular orientation of the capacity for deliberation in the Aristotelian or, better still, the Ignatian sense. This capacity, when shaped by a Christian perspective, should dispose the Christian physician to select among the many particulars of a concrete moral choice those which most closely conform to the virtue of Charity. This entails a certain kind of Christian phronesis, a practical wisdom oriented and motivated by the virtue of Charity to act in a way pleasing to God in any particular situation.

Three aspects of medical moral decisions will serve to illustrate how the Christian virtue of Charity shapes moral choice: (1) in the way the dominant principles of medical ethics are interpreted; (2) in the way the physician-patient relationship is construed; and (3) in the way certain concrete choices in contemporary professional ethics are made.

V. CHARITY AND THE PRINCIPLES OF ETHICS

An agapeistic ethic is by definition a virtue-based ethic. It must, therefore, confront the dilemma of how virtue relates to rules, duties, and principles. This is a prickly problem for any comprehensive philosophy of the moral life. It dates back to the post-medieval departure from virtue ethics exemplified in the overly close conjuctions of canon law with moral theology in the Catholic tradition and with the ascendance of Kant's deontology and Mill's utilitarianism in Anglo-American medical ethics.

The problem is particularly relevant for Christian ethics, for, as Plé points out, "It is not possible to love out of duty, that is, solely for the reason that authority imposes upon me the obligation of loving" ([44], p. 343). To recognize this fact is not to agree with Plé's full indictment of the morality of duty as an "obsessional neurosis" or with the primarily psychologistic line of his argument.

Principles, rules, and duties are as much a reality of the moral life as love and cannot be fully disengaged from it. What may be the essential difference in an agapeistic ethic is that rules, duties, and principles are chosen – or shaped – by Charity, i.e., by whether or not they foster its growth, a fact even Plé admits ([44], p. 344).

The primary principles of medical ethics – beneficence, justice, and autonomy – are ascertainable by human reason without resort to revelation or Sacred Scripture. They enjoy widespread acceptance today even in our morally pluralist society. What the virtue of Charity adds is a special way in which these principles are to be lived and applied in concrete situations. In a Christian moral perspective, Charity "informs" these principles. When they are in conflict, it sorts out resolutions that are in the spirit of Gospel teaching from those that are not.

Each of the principles of medical ethics is thus subject to tests of conformity with sources of moral validation – Sacred Scripture, the tradition or teaching of an official church – not acknowledged by the non-believer. As a result, a Christian or Catholic perspective may impose levels of obligation that, on purely naturalist grounds, are optional or supererogatory. Indeed, the Christian is exhorted unequivocally to perfection in Charity by the Sermon on the Mount. On reason alone such a pursuit can be accounted as unreasonable, unrealistic, or psychologically distressful.

Beneficence – acting for the good of the patient – is the central principle of medical ethics. But beneficence is interpretable at several levels, from mere non-maleficence to heroic sacrifice. Health professionals may differ sharply on precisely what degree of beneficence they consider binding. Some argue that not harming the patient is sufficient, invoking the oft-quoted principle *"primum non nocere"*. Others feel bound to a more positive interpretation, i.e., acting for the good of the patient, not just avoiding harm, thus injecting some degree of altruism. Still others feel impelled to benevolent self-effacement – that is to say, acting for the patient's good even if it means doing so at some personal cost in time, convenience, danger to self or financial loss. Finally, for a very few, like Mother Teresa or Father Damien, beneficence means heroic sacrifice and complete dedication to the needs of the sick and dying.

It can be argued even on purely philosophical grounds that simple non-maleficence, without some degree of self-effacing beneficence is insufficient in medical ethics given the nature of illness, the effects it produces in the patient, and the obligations the physician assumes when she/he offers to heal. I would argue that effacement of self-interest is required even on purely philosophical grounds[37]. This level of beneficence flows directly from the internal morality of medicine and is intrinsic to the traditional concept of a true profession. It is what sets true professions apart from a business, craft, or other occupation.

From a Catholic or Christian perspective, however, benevolent self-effacement is a minimum obligation consistent with the virtue of Charity. Lesser degrees of benevolence and beneficence would be inconsistent with such Scriptural exhortations as the Story of the Good Samaritan, the Sermon on the Mount, or Jesus' own healing acts. On this view medical knowledge is not proprietary or simply a means to make a living. It is a means of service to others, a mission and apostolate, a virtual ministry to those who have a special claim on the whole Christian community – the sick, disabled, poor, or retarded[22].

Thus, for the Christian the practice of medicine is transformed from a profession to a vocation, a means of gaining one's own salvation, assisting in the salvation of others, and witnessing the truth of the Gospel teaching in one's own life. Practicing medicine is inseparable from leading a life that is wholly Christian. The practice of medicine takes on meanings that go beyond even the nobler traditions of the profession. The Christian physician is impelled to act in the interests of the sick even when it may mean exposing himself/herself to danger, loss of time or income, or serious inconvenience. The moral claim of the sick person on the physician exceeds what is expected in a business relationship. Unavailability, inaccessibility, abruptness, condescension, refusal to treat for economic reasons, or fear of contagion are irreconcilable with a Charity-based ethic of medicine.

Sacred Scripture provides no algebraic formula that measures out precisely the degree of self-effacement a particular physician must practice in a specific situation. This will depend on the strength of other obligations to self, family, institutions, other patients, and the like. What is clear is that an agapeistic ethic accepts no easy justification for reducing beneficence to mere non-maleficence or to beneficence which demands nothing of the physician. It makes arguments from exigency, fiscal survival, or adherence to the canons of a competitive environment morally feeble, if not totally unacceptable.

The Christian physician is, in short, called to strive for perfection in Charity even though he or she must fail, given the ineffability of the model he/she must emulate. That Christian physicians and other health professionals do in fact fall short of the benevolent self-effacement Charity requires is obvious. But they should know when they have fallen short. They should strive to come closer always, not as an act of noble self-sacrifice but as an obedient response to a loving God (*Luke* 6: 36) [30].

Similar considerations apply to the other two principles of medical ethics – justice and autonomy.

Seen from a Catholic or Christian perspective justice becomes a charitable justice[21]. It has its origins in God who is just to us and to whom we owe justice. Charitable justice goes beyond the strict rendering to others of what is due to them. It recognizes needs that go beyond duty. It seeks the higher good of the other person in commutative, distributive, and retributive justice.

The Christian is exhorted in the Gospels to "hunger after justice". This means more than fidelity to the natural virtue of justice as taught by Plato, Aristotle or the Stoics. The Christian is called not only to the natural virtues but also to sanctity, to be perfect "as the father is perfect", to cooperate with God in God's work. Charitable justice is not content with rights only. It recognizes claims on us that have no grounding in legal rights but derive from a conception of the human community that enjoins the more fortunate to help the less fortunate whether they "deserve" it or not. There is thus a certain built-in tension between the strictly legal and the Christian senses of justice.

Some of these differences become clear when we examine justice in health care delivery. In a strictly legal sense it is difficult to justify a moral claim by the poor or the sick on the resources of individuals or society. Even more difficult to counter is the argument that the virtuous and the hard working should sacrifice for the poor, the outcasts, sociopaths, alcoholics, or non-complaint in the care of their own health.

Yet it is precisely to these groups that Christians, and specifically Catholic Christians, are expected to exercise a "preferential option"[9]. The recent pastoral letter of the American bishops and the social encyclicals of the Popes since Leo XIII make this clear[53]. When the natural virtue of justice is formed by Charity, it goes beyond a strict calculus of duties and claims and is tempered by compassion. Charitable justice gives freely, lovingly, and without respect to strict accounting, which it leaves to God.

Charitable justice therefore fuses with beneficence in a way incomprehensible on naturalistic interpretations of these virtues. It seeks out those who may not have deserved health care or who are responsible for their own ill-health. It does so because the sick, the poor, the outcast are precisely those to whom Christ himself ministered. No Christian physician can ignore that example and remain authentically Christian.

A specific example of how a Catholic Christian perspective might function is in the selection of a principle of distribution when health care resources are scarce, as in the case of kidneys for transplantation, intensive care beds, or technical procedures of great expense. Distribution theoretically could be on the basis of merit, deserts, societal contribution, needs, first-come-first-served, lottery, or equity[34]. Of these criteria, merit, deserts, societal contribution, and ability to pay are least consistent with the requirements of charitable justice while equity, need, or lottery are more so. Moreover, charitable justice requires that the underlying conditions that lead to rationing choices be eliminated or ameliorated. They cannot be left to the workings of the marketplace as so many suggest today. In a Christian community, health care is an obligation of society, more binding than if it were a legal right.

Autonomy and its accompanying virtue of respect for persons are also to be informed by Charity. Kant grounded autonomy in an *a prioristic* respect for persons. But for the Christian that respect must be grounded in the worth the Creator has given to each life – a worth only God can judge.

The antagonism some ethicists see between autonomy and beneficence is mitigated by a Christian medical ethic[4]. This is not to justify medical paternalism, which is too often confused with beneficence, but to assert that respect for persons is, in itself, a requirement of beneficence. For the Christian, this beneficence is necessary to the virtue of charity. Humans must be free because each has worth, each is accountable to God, each must be free to follow his or her conscience in moral choices – medical or otherwise[1].

Viewed from a Christian and Catholic perspective, however, autonomy is not absolute. The Christian is obliged to use his/her God-given freedom wisely and well. Autonomy is a necessary means to doing the right and the good, to fulfilling the stewardship of our own health. This means refraining from self-destruction by suicide or deleterious life styles, or neglecting needed and appropriate medical care. But if a patient refuses to acknowledge these duties, the physician cannot impose them on him/her. Strong paternalism is uncharitable because freedom to choose and shape one's own life is intrinsic to being human. To ignore it is to violate the very humanity of the patient, a humanity given to him or her by God.

In the same way, the patient and his or her family have an obligation in charity to respect the autonomy of the health professional or institution. The patient cannot, in the name of the absoluteness of autonomy, demand that the physician become the unquestioning instrument of the patient's will. The conscience of the religious physician or hospital cannot be over ridden even if certain practices like abortion, sterilization, discontinuance of food and hydration, or euthanasia are legally sanctioned. The Christian, for example, could not accept the absolutization of patient autonomy and self-governance over life so forcefully promulgated by Judge Compton in his concurring opinion in the *Bouvia* case[5], or as argued by Engelhardt in his recent treatment of the "foundations of bioethics"[10].

In sum, a Catholic Christian perspective shapes the way we interpret and apply the principles of medical ethics in specific ways. Non-believers may interpret them the same way but they need not do so. If they do, they use different reasons. More is rightly expected of those who profess to emulate the example of Jesus' healing. For them the obligations to respect autonomy, justice, and beneficence are at the service of Virtue and Charity. These principles gain their worth not *a prioristically*, but because they express what is necessary to the virtue of Charity.

Whether or not the Christian perspective calls for supererogation if examined on purely philosophical grounds is problematic. The status of supererogation in moral theory – whether it is a separate category or encompassable in Kantian deontology – is still a debatable question[2]. All we need to say at this point is that the range of interpretations of philosophical ethical principles and duties is specified in particular ways in a Charity-based ethic.

Respect for the inviolability of human conscience may bring the Christian physician or nurse into moral conflict with the autonomous decisions of patients, families, other health professionals, the hospital and even the state. In a Charity-based ethics, he/she is obliged to handle the conflict with love and respect for those with whom one disagrees. But on the same principle the Christian health professional cannot cooperate formally or directly with an intrinsically evil act. There is the obligation to decide whether to withdraw respectfully or, if the harm being done – e.g., direct euthanasia, grossly incompetent surgery – is sufficiently great, to intervene directly. One may use those means available in democratic societies – persuasion, ethics committees, and the courts – but not violent means that violate the virtue of Charity.

Needless to say, Christian ethics cannot uncritically accept the current move from substantive to procedural ethics. Such a move is useful and doubtless necessary in a morally pluralistic society. Nonetheless, even in the interests of amicable settlement of moral conflicts, the substance of moral decisions must be defended. This means that a certain tension will exist between secular and Christian ethics.

Likewise, the substantive ethical issues are not resolved if professional ethical codes are revised to fit the needs of morally pluralistic societies. For example, refusing to treat patients with AIDS might be considered "ethical" on the grounds of the AMA principles, which permit physicians to choose whom they will treat. It is hard, however, to defend this view in any authentic interpretation of Christian medical ethics.[3]

VI. CHRISTIAN CHARITY AND PROFESSIONAL PRACTICES

If we move from principles to some concrete moral dilemmas, we can perhaps see more concretely how Charity acts as a principle of discernment and how it orders some of the moral decisions facing health professionals today. We can use as illustrations a few professional practices that have not been declared immoral by the profession as a whole, and which are even accepted, however reluctantly, on grounds of necessity or economic survival. I refer here to a range of practices – some old, some new – which compromise, endanger, or conflict with the best interests of patients. Some examples are : working in for-profit-managed health care systems, medical entrepreneurship in its many forms – investing in and owning health care facilities to which one refers patients, misleading advertising, selling and dispensing medication, refusing to see Medicare or Medicaid patients, charging excessive fees, gatekeeping in its various forms, pay-as-you-go research ([46,47]), cutting corners to contain costs or enhance profits, the many marketing artifices that ensure success in competition and the marketplace, etc. The list of morally marginal practices is spawned by the current commercialization and monetarization of health care as an industry that legitimates the financial motivations of health professionals, administrators, and owners of health care facilities.

These practices are justified on "practical" or economic grounds as means to cost containment and managerial efficiency. Many consider them salubrious to the general welfare, so long as provisions are made to avoid abuses. Even Catholic and other religiously sponsored hospitals and health professionals are enthusiastically embracing these practices.

But it is difficult to justify such practices in any truly agapeistic ethic. The possibilities of submergence of the patient's interest by the financial interests of the health professional or institution, and the downgrading of medicine from a vocation to a business, are all too obvious. Many of these practices have been condemned repeatedly as morally dubious even on non-religious grounds. How much more reprehensible are they when practiced by physicians and institutions that lay claim to the title "Christian"? It is often in the realm of the morally marginal, rather than the frankly immoral, that the more stringent requirements of an agapeistic ethic are most easily discerned.

Today many Christian physicians, nurses, administrators and hospitals justify their compromising the virtue of Charity on grounds of exigency and survival. They thus give proof to the mordant observation of Machiavelli that "a man who wishes to act entirely up to his professions of virtue soon meets with what destroys him among so much that is evil."[4] But is not the challenge of Christian ethics to do precisely what Machiavelli thought impossible?

VII. CHARITY AND THE PHYSICIAN-PATIENT RELATIONSHIP

Similarly, adherence to a Christian and Charity-based ethic shapes the model of the physician- or nurse-patient relationship. Certain of the models now being proposed become morally distasteful if not totally unacceptable. Thus, a genuine Christian ethic would be incompatible with health care as a commercial activity. The idea of the physician as primarily a businessman is inconsistent with the Christian ethic of medicine. Likewise, such an ethic would reject the healing relationship as primarily an exercise in applied biology or as a legal contract for services. Nor could the relationship be construed as paternalistic, or as primarily a means of livelihood, personal profit, or prestige for the physician. Equally incompatible are models which make the physician primarily a government bureaucrat, a proletarian employee of a corporation, or an agent of the state as in totalitarian regimes.

Instead, the model of physician-patient relationships most consistent with a Christian ethic is the covenant – the model in which the physician's promise to help is a binding promise to which he or she pledges fidelity ([28,37]). That promise does not call for total, unquestioning submission to the good as defined by the patient but to the higher levels of charitable beneficence alluded to earlier in this essay. On the Christian view the idea of a profession embraces the higher ideas of a commitment to service subsumed in the idea of a Christian vocation [35].

Parenthetically, the Christian conception of the healing relationship imposes certain obligations on the patients as well. Honesty, compliance with the doctor's regimen, refraining from frivolous, frankly unjust, or injurious legal action and respect for the humanity and moral values of the physician are logical corollaries of a covenantal relationship. A mutuality of respect by physician and patient for the virtue of Charity is therefore essential.

The physician-patient relationship – and equally the nurse-, dentist-, pharmacist-patient relationships – does not, however, call for a monastic devotion to medicine to the exclusion of other obligations to family, self, society or country. It does not deny the fact that in some measures medicine is simultaneously a business, a craft, a science, and a technology. But what an agapeistic ethics does preeminently is to place these differing facets of medical practice into a morally defensible order, recognizing when and to what degree they must yield to the ordering principle of Charity.

This is essentially what it means to say, in medical practice, that Charity is the form of the virtues. Charity acts as a practical principle of discernment and a benchmark against which the Christian measures concretely, here and now, the moral worth of his or her practical decisions. It is often said that the Gospel gives us no categorical guidance, no set rules for resolving all the dilemmas of medical ethics. Apart from the beatitudes, this is so. Manifestly, the Gospel could not anticipate every possible moral dilemma that might arise in the history of mankind. But it gives us something more valuable. It teaches that Charity is the form of all the virtues, that Charity is the ordering principle of discernment in moral choice. And it is very specific in detailing what Charity comprises – all of the concrete examples in Christ's own life; of what he meant in concrete situations by the transcendent ethos of Charity which he preached and taught there on the Mount in full view of the Sea of Galilee and the needy of all the world.

The health professional and institution that profess Christianity must heed that Sermon every day in every encounter with the sick. It is in this sense that the Catholic Christian perspective "sees through" and beyond philosophical medical ethics to the virtue of Charity. Charity becomes an interior principle, as it were, that encompasses the philosophically derivable internal morality of medicine and, without abrogating it, transmutes healing into an act of grace.

Kennedy Institute of Ethics
Georgetown University
Washington, D.C., U.S.A.

NOTES

* I am grateful to the Rev. J. D. Cassidy, O. P. for his helpful criticisms and bibliographic suggestions.
1 See articles by McCormick, Fuchs, Curran and Gustafson in Curran and McCormick (eds.), [9].
2 Others have been more specific about the differences they associate with *Catholic* belief and moral life. See Finnis [13], Hauerwas, [19], and May, [27].
3 Recently the AMA has clarified that Principle VI of its 1980 code does not authorize refusal to treat AIDS patients. *See* Council on Ethical and Judicial Affairs: 1988 'Ethical issues involved in the growing of AIDS Crisis', *Journal of the American Medical Association* **259** (9), 1360–1361.
4 See chapter XV of N. Machiavelli's *The Prince* in *The Great Books, Vol. 23*, Encyclopedia Britannica, Chicago, p. 22.

298 EDMUND D. PELLEGRINO

BIBLIOGRAPHY

1. Abbott, W. (ed.): 1962, *The Documents of Vatican II*, Geoffrey Chapman, London.
2. Baron, M.: 1987, 'Kantian Ethics and Supererogation', *Journal of Philosophy* **84**, (5), 237–262.
3. Bars, H.: 1961, *Faith, Hope and Charity*, trans. P. J. Hepburne Scott, Hawthorn, New York.
4. Beauchamp T. and McCullough, L.: 1984, *Medical Ethics: The Moral Responsibilities of Physicians*, Prentice Hall, Englewood Cliffs, N.J.
5. *Bouvia v. Superior Court of the State of California for the County of Los Angeles*, 225 Cal. Rep. 297.
6. Cahill, L. S.: 1989, 'Theological Medical Morality: A Response to Josef Fuchs', in this volume, pp. 93–102.
7. Congregation for the Doctrine of the Faith: 1987, *Instruction on Respect for Human Life in its Origin and on the Dignity of Procreation*, Vatican City.
8. Crosby, J. F.: 1986, 'The Encounter of God and Man in Moral Obligation', *The New Scholasticism* **60**, (3), 317–355.
9. Curran, C. E. and McCormick, R. (eds.): 1980, *The Distinctiveness of Christian Ethics: Readings in Moral Theology II*, Paulist Press, New York.
10. Engelhardt, H. T.Jr.: 1986, *The Foundations of Bioethics*, Oxford Press, New York.
11. Finnis, J.: 1981, 'Natural Law, Objective Morality and Vatican II', in W. May (ed.) *Principles of Catholic Moral Life*, Franciscan Herald Press, Chicago, pp. 113–149.
12. Fletcher, J.: 1966, *Situation Ethics; The New Morality*, Westminster Press, Philadelphia.
13. Frankena, W. K.: 1963, *Ethics*, Prentice-Hall, New York.
14. Fuchs, J.: 1989 'Catholic' Medical Moral Theology?', in this volume, pp. 83–92.
15. Gilby, T.: 1966, 'Philosophical and Theological Morals' in *Principles of Morality, Summa Theologiae, Vol. 18*, Blackfriars, McGraw-Hill, New York, pp. 147–50.
16. Gilleman, G.: 1959, *The Primacy of Charity in Moral Theology*, trans. W. Ryan and A. Vachon, Newman Press, Westminster Maryland.
17. Guardini, R.: 1954, *The Lord*, Henry Regnery, Chicago.
18. Hallett, G. L.: 1983, *Christian Moral Reasoning*, University of Notre Dame Press, Notre Dame.
19. Hauerwas, S. and MacIntyre, A. (eds.): 1983, *Revisions: Changing Perspectives in Moral Philosophy*, University of Notre Dame Press, Notre Dame.
20. Hawkins, D. J. B.: 1963, '*Christian Ethics*', Hawthorne Books, New York.
21. Hollenback, D.: 1977, 'Modern Catholic Teachings Concerning Justice', in J. C. Haughey (ed.), *The Faith That Does Justice, Woodstock Studies II*, Paulist Press, New York, pp. 207–233.
22. John Paul II: 1984, *Salvifici Doloris*.
23. Jonsen A. R. and Toulmin, S.: 1988 *The Abuse of Casuistry*, University of California Press, Berkely.
24. Krapiec, M.: 1983, *I-Man: Outline of Philosophical Anthropology*, Mariel, New Britain, Conn.
25. Ladd, J.: 1983, 'Internal Morality of Medicine: An Essential Dimension of the Patient-Physician Relationship' in E. Shelp, (ed.), *The Clinical Encounter*, Kluwer Academic Publishers, Dordrecht, Holland, pp. 209–231.

26. Lawler, R. D.: 1986, 'Personalist Ethics', *Proceedings of the American Catholic Philosophical Association* **LX**, 148–155.
27. May, W. E. (ed.): 1981, *Principles of Catholic Moral Life*, Franciscan Herald Press, Chicago.
28. May, W. F.: 1983, *The Physician's Covenant*, Westminster Press, Philadelphia.
29. McCormick, R.: 1985, 'Therapy or Tampering?: The Ethics of Reproductive Technology', *America* **153**, 397–398.
30. McNeill, D., Morrison, D. A., and Nouwen, H. J. M.: 1982, *Compassion, A Reflection on the Christian Life*, Doubleday, New York.
31. Merton, T.: 1985, *Love and Living*, Harcourt, Brace, Jovanovich, New York.
32. O'Brien, T. C.: 1974, 'The *Reditus ad Deum*: A Reflection on the Methodology of St. Thomas' in T. C.O'Brien (trans. and ed.), *Summa Theologiae Vol. 27*, Blackfriars, McGraw-Hill, New York.
33. Oppenheim, F. M. (ed.),: 1986, *The Reasoning Heart*, Georgetown University Press, Washington, D.C.
34. Outka, G.: 1974, 'Social Justice and Equal Access to Health Care', *Journal of Religious Ethics*, **2**, 11–32.
35. Pellegrino, E. D.: 1987, 'Professional Ethics: Moral Decline or Paradigm Shift'. *Religion and Intellectual Life*, **4** (3), 21–39.
36. Pellegrino, E. D.: 1979, 'Toward A Reconstruction of Medical Morality: The Primacy of the Act of Profession and the Fact of Illness', *The Journal of Medicine and Philosophy* **4**, (1), 32–56.
37. Pellegrino, E. D. and Thomasma, D.: 1988, *For the Patient's Good: The Restoration of Beneficence in Health Care*, Oxford, New York.
38. Pinckaers, S.: 1983, 'Autonomie et Heteronomie en Morale Selon S. Thomas D'Aquin' in C. J. P. de Oliveira (ed.), *Autonomie: Dimensions Éthiques de la Liberté: Études d' Éthique Chretienne*, Editions du Cerf, Paris, pp. 104–123.
39. Pinckaers, S.: 1982 'Le Commentaire Du Sermon sur la Montagne par S. Augustin et la Morale de S. Thomas d'Aquin' in L. B. Gillon (ed.), *La Teologoa Morale Nella Storia e Nella Prolematica Attaule Miscellanea*, Massimo, pp. 105–125.
40. Pinckaers, S.: 1984 'La Beatitude dans L'Éthique de S. Thomas' in L. J. Elders and K. Hedwig (eds.), *Studi Tomistici*, Pontificia Accademia, Vatican, pp. 80–94.
41. Pinckaers, S.: 1985, *Les Sources de la Morale Chretienne: Sa Méthode, Son Contenu, Son Histoire*, Éditions du Cerf, Paris.
42. Pinckaers, S.: 1986, *Universalité et Permanence des Lois Morales*, Editions Universitaires, Fribourg.
43. Pittenger, N.: 1981, *Catholic Faith in a Process Perspective*, Orbis, Maryknoll, N.Y.
44. Plé, A.: 1986, 'The Morality of Duty and Obsessional Neurosis' *Cross Currents* **37**, 343–357.
45. Ratzinger, J.: 1980, 'Magisterium of the Church, Faith, Morality', in C. Curran and R. McCormick (eds.), *The Distinctiveness of Christian Ethics: Readings in Moral Theology II:*, Paulist Press, New York.
46. Reade, J. M. and Ratzan, R. M.: 1987, 'Yellow Professionalism: Advertising by Physicians in the Yellow Pages', *New England Journal of Medicine* **316**, (21), 1315–1319.
47. Relman, A. S.: 1987, 'Practicing Medicine in the New Business Climate', *New England Journal of Medicine* **316**, (18), 1150 ff.

48. St. Augustine: 1947, *Faith, Hope and Charity*, trans. L. Arand, Newman Bookshop, New York.
49. St. Augustine: 1948, *The Lord's Sermon on the Mount*, trans. J. Jepson, Newman Press, Westminster, MD.
50. St. Thomas Aquinas: 1974, *Summa Theologiae*, Blackfriars, McGraw-Hill, New York.
51. Sokolowski, R. 1989, 'The Art and Science of Medicine', in this volume, pp. 263–276.
52. Spohn, W.: 1986, 'The Reasoning Heart: An American Approach to Christian Discernment' in F. M. Oppenheim (ed.), *The Reasoning Heart* Georgetown University Press, Washington, D.C., pp. 51–76.
53. U. S. Catholic Conference: 1981, *Catholic Social Teaching and the U.S. Economy: Health and Health Care: A Pastoral Letter of the American Catholic Bishops*.
54. Walgrave, J. H.: 1984, 'The Personal Aspects of St. Thomas' Ethics', in L. Elders and K. Hedwig (eds.), *Studi Tomistici* Pontificia Accademia, Vatican, pp. 202–215.
55. Wren, T. E.: 1985, 'Metaethical Internalism: Can Moral Beliefs Motivate?', *Proceedings of the American Catholic Philosophical Association.*, **59**, 58–80.
56. Wojtyla, K.: 1979, *The Acting Person*, trans. A.Potocki, *Analecta Husserliana*, **10**, Kluwer Academic Publishers, Dordrecht, Holland.

NOTES ON CONTRIBUTORS

Walter J. Burdhardt, S. J., is the editor of *Theological Studies* and is Theologian-in-Residence at Georgetown University, Washington, D.C.

Lisa Sowle Cahill, Ph.D., is Associate Professor in the Theology Department at Boston College, Chestnut Hill, Massachusetts.

Klaus Demmer, M.S.C., is Professor of Moral Theology at the Pontifical Gregorian University in Rome.

Jude P. Dougherty, Ph.D., is Dean and Professor of Philosophy in the School of Philosophy at The Catholic University of America, Washington, D.C.

Josef Fuchs, S. J., is Professor of Moral Theology at the Pontifical Gregorian University in Rome.

John Collins Harvey, M.D., is Professor of Medicine at Georgetown University Medical Center, Washington, D.C.

J. Bryan Hehir is a Senior Research Scholar at the Kennedy Institute of Ethics, Georgetown University, Washington, D.C. and the Secretary of the Department of Social Development and World Peace at the U. S. Catholic Conference.

Monika Hellwig, Ph.D., is Professor of Systematic Theology at Georgetown University, Washington, D.C. and is currently President of the Catholic Theological Society of America.

Gerard J. Hughes, S. J., is Head of the Philosophy Department at Heythrop College, University of London, London.

John P. Langan, S. J., is the Rose F. Kennedy Professor of Christian Ethics at the Kennedy Institute of Ethics, and a Senior Fellow at the Woodstock Theological Center, Georgetown University, Washington, D.C.

Robert F. Leavitt, S. S., is President and Rector of St. Mary's Seminary and University, Baltimore, Maryland.

His Eminence, Carlo Maria Martini, S. J., is the Archbishop of Milan.

William E. May, Ph.D., is Ordinary Professor of Moral Theology in the

301

Department of Theology at The Catholic University of America, Washington, D.C.

Richard A. McCormick, S. J., is the John A. O'Brien Professor of Christian Ethics in the Department of Theology at the University of Notre Dame, Notre Dame, Indiana.

E. D. Pellegrino, M.D., is Director of the Kennedy Institute of Ethics, and is the John Carroll Professor of Medicine and Medical Humanities at Georgetown University, Washington, D.C.

Bruno Schüller, S. J., is Professor of Moral Theology at Westfalische Wilhelms-Universität in Munster.

Robert S. Sokolowski, Ph.D., is Professor of Philosophy at The Catholic University of America, Washington, D.C.

David C. Thomasma, Ph.D., holds the Michael I. English, S. J., Chair of Medical Ethics and is Director of the Medical Humanities Program at the Stritch School of Medicine, Loyola University Medical Center, Maywood, Illinois.

William A. Wallace, O. P., is Professor of Philosophy and History of Science at the Catholic University of America, Washington, D.C.

John H. Wright, S. J., is Professor of Systematic Theology at the Jesuit School of Theology, Berkeley, California.

INDEX

abortion 1, 50, 51, 58, 59, 90, 119, 132,
149–150, 153, 157, 162, 176–179, 207,
212–215, 217, 219–220, 224, 226
abundance (principle of) 118–119
agape 65, 155
agapeistic ethics 277–297
Albertus Magnus 28
altruism 290
American Medical Association (AMA)
156, 294, 297
Antonelli, J. 134, 143n
Antoninus, St. 3, 134, 143n
apatheia 89
Aquinas, St. Thomas 17, 23–51, 52n, 79,
93, 94, 97, 133, 135, 138, 140, 143,
171, 197, 231, 234, 245, 256–257,
284–287
Aristotle 3, 15, 16, 23–51, 52n, 63, 66,
71, 93, 185, 234, 256–257
artificial insemination 64
Ashley, B. M. 52n
Augustine, St. 79, 132, 133n, 158, 171,
231, 233, 256, 286
autonomy (principle of) 16, 108, 111, 239,
251, 291–293
Averroes 28, 256

Balthasar, H. U. von 76, 77n
Barry, B. 62, 77n
Barth, K. 108
Bauer, W. 232, 253n
Beauchamp, T. L. 61, 77n
Bell, D. 252, 253n
beneficence (principle of) 279, 281, 289,
292–293
Bernardin, J. Cardinal 98, 157, 166n
biomedical ethics 1, 15, 23
catholic tradition and 1–5, 83–92, 93,
129–143

deontological approach 2–3, 61, 289,
294
natural norms and 23–53, 47–51, 55–58
procedural approach 3
professional philosophers and 2
public policy and 217–220, 224
scripture and 96, 99, 124, 136–137, 291
teleological approach 61, 96, 99
theological approach 2, 83–92, 93–101,
103–122, 136, 141, 271–274
utilitarian approach 2, 289
Blondel, M. 239, 253n
Boethius 257
Bouvia vs. Superior Court 293, 298n
Brandt, R. 61, 77n

Cahill, L. 135, 137–140, 143, 278, 280
Capellman, C. 134, 144n, 171
casuistry 1, 9, 14, 94, 133, 278
Catholic Health Association 218, 221n,
228, 229n
causality 25–26, 51, 120
charity (virtue of) 3, 4, 274, 277–297
Childress, J. 61
church-state relationship 206, 210
Civiltà cattolica 147
Clement of Alexandria 132, 232
Collins, R. F. 95, 102n
conception 48, 58, 76, 150–151
delayed hominization 48, 50, 79
immediate hominization 48, 51
Connery, J. 156, 166n
constitution (U.S.) 205–206
contraception 1, 50, 64, 132, 152, 159,
161, 175
covenant 296
creation 26, 36–40, 80, 120–121
creationism 56, 58
Crosby, J. 288, 298n

The Philosophy and Medicine Book Series

Editors

H. Tristram Engelhardt, Jr. and Stuart F. Spicker